the deeper genome

John Parrington is an Associate Professor in Molecular and Cellular Pharmacology at the University of Oxford, and a Tutorial Fellow in Medicine at Worcester College, Oxford. He has published over 80 peer-reviewed articles in science journals including *Nature, Current Biology, Journal of Cell Biology, Journal of Clinical Investigation, Development, Developmental Biology*, and *Human Reproduction*. He has extensive experience writing popular science, having published articles in *The Guardian, New Scientist, Chemistry World*, and *The Biologist*. As a British Science Association Media Fellow he worked as a science journalist at *The Times* for 7 weeks where he published 22 articles. He has also written science reports for the public for the Wellcome Trust, British Council, and Royal Society. http://www.johnparrington.com

'an excellent synthesis of what is currently understood about the human genome. It corrects claims of omniscience and shows just how wrong informed predictions turned out to be'

Hamilton McMillan, *The Guardian*

'recommend . . . for a stereoscopic view'

Peter Forbes, *The Independent*

'a good primer on the subtlety and complexity of the genome, especially the human genome, new facets of which emerge on a regular basis from labs around the world.'

Bob Grant, *The Scientist*

'Overall, this is a faithful, engaging portrait of the twenty-first-century genome'

Nathaniel Comfort, *Nature*

'a great read that definitely imparts knowledge in an entertaining fashion and connects the almost 99 percent of the genome that is not protein coding to all sorts of interesting questions. We highly recommend it.'

Dov Greenbaum and Mark Gerstein, *Cell*

'This informative, highly readable book addresses scientists' current understanding of noncoding DNA. . . . All in all, this is a valuable book for anyone wishing to explore the newest discoveries, and the implications of these discoveries, in a rapidly expanding field. Highly recommended.'

CHOICE

'provides an elegant, accessible account of the profound and unexpected complexities of the human genome, and shows how many ideas developed in the 20th century are being overturned.'

Clare Ainsworth, *New Scientist*

the deeper genome

Why there is more to the human genome than meets the eye

JOHN PARRINGTON

OXFORD
UNIVERSITY PRESS

OXFORD
UNIVERSITY PRESS

Great Clarendon Street, Oxford, OX2 6DP,
United Kingdom

Oxford University Press is a department of the University of Oxford.
It furthers the University's objective of excellence in research, scholarship,
and education by publishing worldwide. Oxford is a registered trade mark of
Oxford University Press in the UK and in certain other countries

First Edition published 2015
First published in paperback 2016

Impression: 1

Published in the United States of America by Oxford University Press
198 Madison Avenue, New York, NY 10016, United States of America

British Library Cataloguing in Publication Data
Data available

Library of Congress Cataloging in Publication Data
Data available

ISBN 978–0–19–968873–9 (Hbk.)
ISBN 978–0–19–968874–6 (Pbk.)

Printed in Great Britain by
Clays Ltd, St Ives plc

ACKNOWLEDGEMENTS

I would like to thank a number of people who have helped bring this book to fruition. I owe particular thanks to Latha Menon, my editor at Oxford University Press, who was both firm in her suggestions about where the text needed modifying, and encouraging where she felt it did not. I would also like to thank Emma Ma and Jenny Nugee of the OUP editorial team, for their help on a multitude of practical matters, and Elizabeth Stone at Bourchier Limited for her meticulous copy-editing of the book. I gained some very valuable insights and suggestions for modifications to the text from a number of people who read my original proposal and various drafts of the book, namely Guida Ruas and Martin Empson, together with four anonymous reviewers. I also owe many thanks to Anthony Morgan for producing the photo for the book cover. For their excellent assistance with marketing and publicity I would like to thank Phil Henderson and Kate Farquhar-Thomson of OUP, as well as Jonathan Wood of the Oxford University Press Office. I would also like to thank Kate Gilks of OUP and Andrew Hawkey for their skill and expertise in proof-reading and compiling the index. I am very grateful to friends and colleagues who have indulged my many queries and speculations about matters relating to the genome during the writing of this book, as well as providing very helpful feedback and suggestions. Finally, I owe special thanks to my family, who have provided me with love throughout the writing and production of this book, and put up with the many hours spent researching and writing when it cut into our time spent together as a family.

CONTENTS

INTRODUCTION

How the Genome Lost Its Junk

'Sit down before fact as a little child, be prepared to give up every preconceived notion, follow humbly wherever and to whatever abysses nature leads,
or you shall learn nothing.'   *Thomas Huxley*

'What is a scientist after all? It is a curious person looking through a keyhole,
the keyhole of nature, trying to know what's going on.'   *Jacques Cousteau*

It was on the morning of 5 September 2012 that I first heard about the death of
'junk' DNA. I was sitting at a desk at *The Times* newspaper in London; to one side,
through huge windows, I could see the Thames, Tower Bridge—which all summer had been sporting the Olympic and Paralympic symbols—and beyond that
the Shard, the London Eye, and other famous landmarks. Above me, in the open-
plan building occupied by Rupert Murdoch's News International company, was
the floor occupied by the *Sun*, with its huge, framed past front pages with headlines like 'Up Yours Delors!' and 'Sling Your Hook!', references to European
Commission President Jacques Delors and radical Muslim cleric Abu Hamza,
respectively. At the time the *Sun* was embroiled in a major investigation into its
alleged use of illegal phone tapping.[1] Although it was 9.30 a.m. the offices were still
largely empty; the deceptive lack of activity was contradicted, however, by the influx
of messages in my e-mail inbox from other journalists, pitching ideas to the editors
for the day's stories even as they travelled to work by the Underground or rail.

Although I'd been working at *The Times* for over a month, my position funded
by a British Science Association Media Fellowship,[2] I still felt a bit of an imposter,
perhaps because the day-to-day activities of being a journalist were so different

compared to my normal role as a biologist and lecturer at Oxford University. One particular difference was the tempo; while in my regular job I may spend months, even years, gathering data for a study and presenting it for publication, submitting the manuscript to a journal, and then spending more time battling with anonymous reviewers who can either damn the whole study with a dismissive word or demand further data, here the pace of publication was very different.

So a typical day at *The Times* began by scouring Eureka Alert and other websites that gather together the latest press releases, funding announcements, and other news from the world of science.[3] This would form the basis of my day's pitch to the news editors, which typically would consist of two, maybe three, stories I thought might compete with other news from the world of politics, economics, sport, and scandal. After anxiously waiting while the editors had their mid-morning meeting, I would hopefully get the go-ahead to write 600 words on one topic, 400 on another, all to be submitted to the news desk by 3 or 4 p.m. to have any chance of making the printed paper. Around me, kick-started into life by similar demands, the office was now a whirring hub of activity as everything became subsumed towards a central goal—the production of the next day's news. If I had written well, and, as important, proved lucky against competing news items, I might see one or two of my articles online by early evening. However, the real test of how well I was doing would be seeing a piece that I'd written appear in next day's print edition. And then, like rubbing clean a slate, the next day kicked off exactly the same way.

This morning, however, it was clear something odd was afoot. Over a dozen different press releases had appeared on Eureka Alert, all from different research institutions, but all mentioning ENCODE—an acronym for ENCyclopedia of DNA Elements. As I read further, I learnt the reason for this sudden burst of information: ENCODE was the culmination of almost a decade's research involving 442 scientists from 32 institutions and costing $288 million.[4] And its claims seemed as big as its budget. So while the original Human Genome Project provided the sequence of letters that make up the DNA code, ENCODE appeared to have gone substantially further and told us what all these different letters actually do. Perhaps most exciting was its claim to have solved one of the biggest conundrums in biology: this is the fact that our genes, which supposedly define us as a species, but also distinguish you or I or anyone else on the planet from each other, make up only 2 per cent of our DNA. The other 98 per cent had been written off as 'junk'; however, this raised the question of why our cells should spend vital energy replicating and storing something with no function. The existence of so much junk DNA had also featured heavily in debates between evolutionists and creationists, for why would any creator design a genome in which only 2 per cent actually works?

Now, as I read though, I found that ENCODE's new findings, all synchronized to appear simultaneously in 30 linked publications, had a new and excitingly different take on this matter. By scanning through the whole genome rather than just the genes, and using multiple, cutting-edge approaches to measure biochemical activity, ENCODE had come to the startling conclusion that, far from being junk, as much as 80 per cent of these disregarded parts of the genome had an important function. Indeed, for Ewan Birney of the European Molecular Biology Laboratory near Cambridge, the charismatic spokesperson of the project, this was probably an underestimate, since it was 'likely that 80 percent will go to 100 percent. We don't really have any large chunks of redundant DNA'.[5] The path-breaking nature of the project was emphasized by another ENCODE researcher, John Stamatoyan-nopoulos of Washington University in Seattle, who predicted that the findings would 'change the way a lot of concepts are written about and presented in textbooks'.[6] Perhaps most excitingly for a general audience, ENCODE also claimed that its findings were casting important new light on links between the genome and common diseases such as heart disease, diabetes, auto-immune conditions, and mental disorders like schizophrenia.[6]

Clearly, this seemed like big science at its best, and, as such, I co-wrote a story with Tom Whipple for The Times, in which we spelled out the study's implications. It appeared in the following day's paper entitled 'Rummage through "junk" DNA finds vital material'.[7] Similar positive assessments of the new findings appeared in media outlets across the world, all of which repeated the project's main conclusion that the idea of 'junk' DNA had been overturned by the discovery that as much as 80 per cent of the genome had an important function.[8,9,10] This was also the message from serious science journals such as Nature and Science, with the latter headlining the discovery 'ENCODE project writes eulogy for junk DNA'.[11]

By the end of my six-week placement at The Times, I'd published a total of twenty-two stories and features, on topics ranging from why chocolate is addictive, the discovery of the oldest tooth filling in history, and whether becoming a eunuch would make men live longer, to the burning question of whether a potential super-volcano is lurking under the city of Naples.[12] But the story that most resonated for me on a personal level was the ENCODE findings. I resolved to find out more about their implications, not just for my own research, but also with regard to the much bigger question of how our genomes define us both as a species and as individuals. Most intriguing was the controversy that erupted a little while after the ENCODE findings were published. So an article published in the journal *Genome Biology and Evolution* in February 2013, attacked the findings in a vitriolic tone not normally associated with scientific debate, or at least not in the pages of an academic journal.[13] According to the article, the claims of ENCODE were 'absurd', its statistics 'horrible', and it was 'the work of people who know nothing about evolutionary biology'. And in a subsequent interview, lead author, Dan Graur of Houston University, said 'this is not the work of scientists. This is the work of a group of badly trained technicians.'[14] A central criticism was that ENCODE researchers had confused activity with functionality. 'Just because a piece of DNA has biological activity does not mean it has an important function in a cell,' said Graur. 'Most of the human genome is devoid of function and these people are wrong to say otherwise.'[14]

In contrast, there was the view of John Mattick of the Garvan Institute of Medical Research in Sydney, who argued that the ENCODE leaders were, if anything, too conservative in their claims, and that the findings showed 'we have misunderstood the nature of genetic programming for the past 50 years'.[15] Another proponent of this view, Evelyn Fox Keller of the Massachusetts Institute of Technology, believes recent 'genomic science has changed the very meaning of the term, turning the genome into an entity far richer, more complex, and more powerful—simultaneously both more and less—than the pre-genomic genome, in ways that require us to rework our understanding of the relation between genes, genomes and genetics'.[16]

So who is right? I resolved to find out, and this book is partly a result of that quest. However, while my interest in writing this book began with ENCODE, it has subsequently grown to encompass a much wider field of enquiry, all relating to the topic—how do our genomes make us human? This is a question that often comes up in the lectures and tutorials I give to medical and biology students at Oxford University in which we discuss the genetics of disease; for instance, why are some people more susceptible to certain disorders than others? But it also flows from a long-standing personal interest in what distinguishes humans as a species, but also as individuals. Unfortunately, for some time now I've been dissatisfied with the available explanations as to how our human genomes work, on the one hand to distinguish us from other species on the planet, and on the other to create the unique mix of personality, capabilities, needs, desires, and susceptibility to illnesses and disorders, that define us as individuals.

As a child, I remember being fascinated by my parents' copy of The Naked Ape, by Desmond Morris. This book became a publishing sensation in the 1970s with its claim that modern human behaviour and society was largely rooted in instincts that had evolved in the Stone Age.[17] There was a problem though, in that Morris's take on prehistoric life was about as accurate as the 1950s cartoon The Flintstones, or the '60s film One Million Years BC starring Raquel Welch. However, what The Naked Ape lacked in authenticity was more than compensated by its numerous references to sex, which, to someone just reaching puberty, were almost as alluring as Welch's animal skin bikini. For a teenage boy just starting to worry about my attractiveness to the opposite sex, being told that humans not only have the largest brain compared to body size, but also the largest penis, making them the 'sexiest primate alive',[18] was sweet music to my fragile ego. Meanwhile, learning that humans' fleshy ear lobes, unique to our species, are erogenous zones, or that women's breasts are an important sexual signalling device rather than simply providing milk for babies,[19] was definitely something for my newly hormone-stimulated brain to chew over. And finally, Morris's claim that monogamy evolved so that men out hunting could trust their mates were not having sex with other men, and that human 'nakedness' helped intensify pair-bonding by increasing sensory pleasures,[19] were all food for thought to a teenage mind.

Unfortunately, as scientific fact such claims were about as substantial as Welch's bikini, although they could possibly be excused by a lack of understanding of human evolution and its molecular and cellular basis at this time. Yet what is surprising is how little many 'biological' explanations of human behaviour and society have changed since the 1970s. Take, for example, a recent book about human culture by Mark Pagel of Reading University, which moves in a few pages from self-sacrifice in amoeboid slime moulds to what Pagel calls a 'helping gene' that codes 'for an emotion that disposes people to be friendly'.[20] This sounds quite nice except this helping gene's influence only extends to people of the same nation; towards other national groups it becomes the 'jingoism' or 'xenophobia' gene. The author ends by saying 'next time you feel that warm nationalistic pride at the sound of your national anthem or the news of one of your country's soldiers' valour, think of the amoebae!'[21]

One problem with this argument is its assumption that all members of a nation state behave in a similarly patriotic manner. So it fails to explain why millions of people in Britain opposed their government and marched against the recent wars in Iraq and Afghanistan.[22] But a more fundamental difficulty is that the proposed 'gene', which somehow manages to combine both nice and nasty characteristics, is a complete figment of the author's imagination. And, lest this be seen as an isolated incident, I could point to a range of other examples in which single genes are said to determine intelligence, personality, and even men's supposed unwillingness to do the ironing, without scientific evidence to back up such claims. In fact, for all the lip service paid to genetics in such accounts, the 'gene' here might as well be made of green cheese given the lack of any real attempt to engage with actual molecular mechanisms rooted in the real genome.

Actually, this is not quite true. The claim that homosexuality is due to a 'gay' gene, which made headlines across the world in 2003, was based on a study published in the prestigious journal *Science*. Evidence was presented that gay men had specific differences in a region on the X chromosome, Xq28, that was claimed to be linked to their homosexuality, and passed down through the mother.[23] What followed was a huge debate about the implications of the discovery. In the gay community itself reactions ranged from fears that screening programmes might identify and abort 'gay foetuses' on the basis of their possession of the gene, to those who thought the discovery would scientifically 'legitimize' homosexuality and therefore help end gay oppression, although this ignores the fact that the clear biological basis of skin colour has not prevented oppression based on this difference.[24] Such was the publicity around the discovery that a T-shirt with the slogan 'Xq28—thanks for the genes, Mom!' became a popular item in many gay bookstores. Missing from much of the coverage, however, was that what had been discovered was not an actual gene but merely an association with a DNA region on the X chromosome. And two decades later, the authors have failed to identify such a gene, while attempts to reproduce the findings by others have been equally negative, leading to suspicions that the original 'discovery' was just a statistical artefact.[25]

Such failed attempts aside, one undoubted reason for the popularity of accounts of human behaviour and society that make no attempt to engage with actual molecular mechanisms, is that they tell entertaining 'just-so' stories about human evolution without having to bear scrutiny as to whether such stories are actually true. In this sense, as science journalist Tim Radford has noted, the 'gene' here is not so much a real object as a 'metaphor, an analogy, an "as if", a useful way of thinking about how behaviours, strategies and responses might have emerged'.[26] Now while I'm all for metaphors as a way of making complicated scientific concepts comprehensible, a problem arises when these get in the way of a true understanding of the material basis of nature. Taking an example from the physical sciences, the initial model of the atom proposed by Ernest Rutherford in 1911 pictured it as a miniature solar system, with electrons orbiting the nucleus just as our own planet orbits the Sun; however, subsequent studies showed that this metaphor was far too simplistic. Surprisingly, in the biological sciences far too many popular accounts are still anchored in an old-fashioned view of genes that either sees them as abstract units with no material form, or if they do acknowledge this link, subscribe to the crude picture of genes as 'beads on a string', discrete and isolated entities on each chromosome.

In contrast, while in this book there will be plenty of speculation about the link between genes and what it means to be human, my aim will be to make sure this is always backed up with evidence of real molecular mechanisms based on the most cutting-edge studies of the genome. Here, though, we face a problem. While key individuals involved in the Human Genome Project, such as Sir John Sulston of the Sanger Institute near Cambridge, promised that, 'for the first time we are going to hold in our hands the set of instructions to make a human being',[27] and British Science Minister Lord Sainsbury said 'we now have the possibility of achieving all we ever hoped for from medicine',[27] the reality has been rather different. So when scientists sought to use the genome to identify links between differences in the DNA code and common disorders like heart disease, diabetes, and mental conditions like schizophrenia, bipolar disorder, and autism, the problem was not finding genetic links to these disorders but rather the astounding number of these.[28,29] Instead of identifying just a few strong genetic links with each condition, as had been commonly predicted, scores or even hundreds of such links have been made, with each only apparently contributing a tiny amount to the chance of succumbing to these disorders. Such findings seem to mock the idea that we can find meaningful, and useful, links between our genomes and such conditions, and if true for disease, surely it must be more so for other aspects of being human, such as our unique personalities and capabilities. This has led one critic of the genome project—neuroscientist Steven Rose of the Open University—to argue that 'they said this was the greatest achievement since landing a man on the moon. One even said it ranked with the discovery of the wheel. And yet none of this cornucopia of benefits has come out of it.'[30] Another critic, bioethicist Tom Shakespeare of Newcastle University, has said 'we share 51 percent of our genes with yeast and 98 percent with chimpanzees—it is not genetics that makes us human'.[27]

In this book, my aim is to find a middle way between the view that the complexities of the human condition can be reduced to simple, hypothetical 'genes' without any mechanistic underpinning, and the opposite view that sees things as far more complex, yet rejects the idea that we have learned anything useful from the genome project, about both the diseases that afflict us as a species, and what it means to be human. In so doing, I will be drawing on many years of experience studying genes and how they function. In my quest to understand how cellular signals regulate important bodily processes, I have isolated and characterized novel genes and studied their functional properties in a test tube and in cultured cells, but also what happens when their activity is inhibited or altered in a living animal. This has allowed me to appreciate the power, but also limitations, of the so-called 'reductionist' approach to biology.

Such an approach aims to understand complex biological systems by dissecting them into their constituent parts, or as Francis Crick, co-discover of the DNA double helix put it, 'to explain all biology in terms of physics and chemistry'.[31] The power of this approach was demonstrated to me when my colleagues and I used it to show that a single gene codes for a protein in the sperm which triggers the chemical signal in the egg that stimulates embryo development.[32,33] However, while studying the role of genes in the living organism, I have also increasingly come to realize the complexity of their behaviour. A common method for studying how genes work and their function within the body is to breed animals in which the action of a particular gene is inhibited by genetic engineering, thus creating a 'knockout' mouse.[34] Such animals have become very important 'models' of human disease. Yet, surprisingly, in many cases abolishing a gene's activity has little effect on the whole organism, or leads to opposite or very different effects than predicted based on its properties in a test tube or in cells in a culture dish. Such unexpected findings are often said to be due to 'compensation' by other genes during embryo development but this is really just a hand-waving gesture to convey the fact we often know very little about why particular genomic manipulations have unforeseen effects.[31] What these findings do suggest is that while isolating the effects of one gene from others in the body can lead to important insights, ultimately, gene action can only be properly understood as part of a wider whole. And if true for a mouse how much more so for humans, with our complex behaviour and culture driven by social innovation as much as by biology.

So does this mean that attempts to find genetic links to the complexities of human behaviour and society are doomed to failure? In this book I intend to show this is far from the case, but I also want to challenge some long-held assumptions in biology: one being the idea that genes can be treated in isolation, and also the very definition of what we mean by a gene. To do this I will not only explore what the ENCODE findings have to tell us about this question, but also investigate what I believe is a more general shift taking place in our perception of the genome and how it works. Importantly, this shift is based on new technologies that mean that, rather than studying single genes in isolation as previously, we can now observe changes in the activity of the genome as a whole.[35] This analysis can extend both to a whole organ like the human brain, but also allows us to study how genes are switched on and off, in real time, in a living cell, something undreamed of only a few years ago.[36]

Based on the findings emerging from the use of such technologies, I will look at important new developments like the increasing recognition that, far from simply being a linear code, the genome only really makes sense as a 3D entity.[37] Moreover, this 3D entity dynamically changes in response to signals originating both from within, and outside, the cell. Another important development is the recognition that RNA, DNA's chemical cousin, plays a far more important role in the cell and organism than previously thought.[38] So instead of simply being a messenger between DNA and proteins—the building blocks of the body—RNA is proving to have a multitude of other key roles, and on a much vaster scale than could have been imagined. Finally, new evidence is emerging that, far from being a fixed DNA 'blueprint', the genome proper is a complex entity that includes proteins, and both the DNA and these proteins can be chemically modified in a far more rapid, and reversible, fashion than suspected.[39] This makes the genome exquisitely sensitive to signals from the environment, and challenges the idea that life is merely a one-way flow of information from DNA to organism. Perhaps most surprisingly, the genome's status as a structurally stable unit is being called into question, with evidence that certain genomic elements have an ability to move about, sometimes to the detriment of normal cellular function, but also acting as a new source of genome function.[40,41,42]

Excitingly, while the significance of such phenomena for long-term evolutionary change remains controversial, there is increasing evidence that these newly recognized features of the genome may have played a fundamental role in the emergence of Homo sapiens as a unique species with self-conscious awareness and the power to transform its environment in a way that sets it apart from all other life forms on the planet.[43] This focus on human beings will be an important theme of this book, for although I will show how studies on organisms ranging from the humble bacterium to our closest living animal cousins, chimpanzees, have transformed our understanding of human biology, ultimately it is with our own species that I will be most concerned. And I will be aided in this task by the fact we can now study the genomes not just of living primates, but also extinct proto-human species like Neanderthals.[44,45] In addition, it is becoming increasingly feasible to determine the complete DNA sequence of genomes, as well as chemical modifications of this DNA and its associated proteins, from large numbers of living human beings.[46]

Before tackling these important new reconsiderations about the genome though, I first want to take us back in time to look at how scientists came to understand how living things appeared on Earth, what led to the diversity of life we see around us, and how genes and genomes mediate this process. In so doing we will reach back into the lives and times of famous scientists like Darwin and Mendel, but also lesser known figures whose theories and findings have nevertheless enriched our view of the cell and organism. Having thus developed a secure foundation based on what, until recently, constituted the 'orthodoxy' in this area of science, we will examine how this orthodox view is currently being challenged. Using such new information we will seek to understand what this can tell us about abnormalities of the human condition, not just 'single-gene' disorders, but also more common disturbances, such as heart disease and diabetes, and also mental conditions like schizophrenia and bipolar disorder, that afflict millions of people across the world. In particular, we will investigate whether the genetic complexity that appears to underlie these disorders mean that attempts to identify new methods of diagnosis and treatment are fatally flawed, or if there is a path to a new understanding of these conditions despite this complexity. Finally, we will explore how such new understanding of the genome is being used to address a key remaining question for humanity, namely what is so special about our own species that led us to such a primary position on Earth. With all that in mind, it's time to begin our quest. But first, a personal question—do you feel lucky?

1

THE INHERITORS

'Every individual alive today, even the very highest, is to be derived in an unbroken line from the first and lowest forms.'   *August Weismann*

'If we didn't have genetic mutations, we wouldn't have us. You need error to open the door to the adjacent possible.'   *Steven Johnson*

Do you ever feel you could use a little extra luck? Most of us can remember missed opportunities when a helping hand from chance wouldn't have gone amiss. Of course, winners in life are often said to make their own luck, but the popularity of lotteries across the world is proof of the hope that great fortune might nevertheless turn up out of the blue with minimal effort. Unfortunately, at odds of 14 million to one, you're four times more likely to be killed by a lightning strike, and seven times more likely to die falling out of bed than become a lottery multi-millionaire.[47] But what if I told you that you're already a winner at odds that, in comparison to a lottery win, would make the latter seem as certain as the sun rising each morning?[48] To calculate just how lucky you are, first consider the chance that accompanied your mother and father meeting and deciding to have a child, estimated at one in 20,000. Then there's the good fortune that, of the four trillion sperm a man generates in his lifetime, and the 100,000 eggs a woman produces, the pair that gave rise to you happened to come together. But really we're only getting started when we consider the unlikeliness of your existence. For we must also consider an even more implausible chain of events, namely that each of your ancestors lived to reproductive age throughout the 3.7 billion years since life began on Earth. As such, the unlikeliness of your birth comes to about 1 in $10^{2,685,000}$, or putting it another way, it's as unlikely as the people in central London getting together, each with a trillion-sided dice, and all rolling the same number.

All of which means you are extraordinarily lucky to be here. But before you get too carried away, you should also know you share similar good fortune with all the other living things on Earth. That's not just the other seven billion human beings, but also the nine million other species on our planet. I'm not even going to try and calculate the total number of organisms on Earth, but the fact that in your guts alone there are 100 trillion bacteria, should give a sense of the scale we're talking about. And yet, like you, each individual organism on the planet came into being through an extraordinarily lucky set of events. Yet there's something even more fundamental than good fortune that you share with all these organisms, and that's a common ancestry. From the mighty elephant to the minuscule flu virus, all of life is, in a sense, our cousins.

This view of life—that undirected blind chance led to each individual organism being alive today on Earth, and also that we share a common ancestry—was famously proposed by Charles Darwin in *The Origin of Species* in 1859.[49] Following his trip around the world on the HMS *Beagle*, and stimulated by the diverse life forms he had seen on the trip, Darwin concluded that there was no necessary requirement for a supernatural creator to have produced the multitude of species on our planet—instead this could be explained by a driving force he termed 'natural selection'. This required both that populations of species varied in their size, shape, and capabilities, and new environmental pressures acted upon these variants to ensure 'survival of the fittest'. Although this phrase conjures up the image of nature as 'red in tooth and claw', Darwin was careful to point out that although 'two canine animals, in a time of dearth, may be truly said to struggle with each other which shall get food and live', the fight to survive is equally valid for 'a plant on the edge of a desert [struggling] for life against the drought'.[50] Another important aspect of the theory is that for such survival to have any consequence for future generations, survivors must pass on their attributes to their offspring, who then represent a more significant proportion of the population. To see how natural selection works, consider giraffes: while their ancestors had shorter necks, a few animals with slightly longer necks would have had a survival advantage in being able to access the highest leaves on the trees. Through selection over generations, these variants came to predominate. Eventually, such differences can lead to the birth of a new species.

In fact, Darwin was not the only person who had this crucial insight—so did Alfred Russel Wallace. Unlike the wealthy, university-educated Darwin, Wallace began working for a living at 13 years old, first as a builder's apprentice, later by collecting biological specimens and selling them to collectors. Wallace was also a socialist who continued promoting his radical views until his death at the ripe old age of 90.[51] However, despite these differences, Wallace reached the same conclusions about the evolutionary process as Darwin through a remarkably similar route. Importantly, he had the same crucial exposure to an extraordinary number of species and their variants during his travels around South America and what is now Indonesia, as Darwin had on the HMS *Beagle*. Moreover, Wallace arrived at the idea of a struggle for existence being the driving force for evolution after reading Thomas Malthus's *An Essay on the Principle of Population*, exactly as Darwin had years earlier. In 1798 Malthus proposed that famine and disease were an inevitable feature of human society, since, while food supply only increased linearly, populations grew in an explosive, exponential fashion. In particular, he believed that the 'lower orders' were primarily to blame in the latter respect, being too inclined to have children.[52] Many critics, then and subsequently, have pointed to flaws in Malthus's reasoning, such as the fact he ignored the likelihood of technological advances in food production.[52] An additional flaw was his disregard for the possibility of birth control, which, ironically, he detested.[52]

However, as a stimulus to the idea of natural selection, Malthus's arguments were central to the development of both Darwin's and Wallace's thought. In Wallace's case, it was in 1858, while fighting a malarial fever in Ternate, Indonesia, that he recalled Malthus's arguments and realized a struggle for scarce resources could provide a mechanism for evolution.[53] In the mid-nineteenth century it was still highly risky to advocate a view of life's origins that left no requirement for God. Although Darwin developed his own version of natural selection as early as 1838, he held back from publishing this, partly out of an obsessive desire to work out every last little theoretical detail, but also out of fear of how going public would affect his respectable position in society.[54] Wallace had no such qualms, and in June 1858 he wrote to Darwin with an outline of his new idea, and a request to review his theory and then send it to the geologist Charles Lyell, who, ironically, had been secretly urging Darwin to publish his work.[53] For Darwin, the letter arrived like a bombshell, for, as he observed to Lyell, 'if Wallace had my manuscript sketch written out in 1842, he could not have made a better short abstract!... So all my originality, whatever it may amount to, will be smashed.'[55]

Instead, a compromise was arranged by Lyell and others in the scientific establishment, whereby both men's views on the subject were presented at a meeting of the Linnean Society in London in July 1858.[54] Neither Darwin nor Wallace attended this, the former being ill and the latter still in Indonesia, and the meeting drew surprisingly little attention at the time. Instead, it was the bestselling *Origin of Species*, published in November 1859, that both introduced the theory to a wider public, and ensured its primary association with Darwin. However, another reason for the relative lack of public recognition of Wallace's contribution may also be his unwillingness to follow the theory through to its logical conclusion and apply it to the origin of human consciousness. So while Darwin did this in 1871 in *The Descent of Man*, where he proposed that humans are 'descended from a hairy, tailed quadruped, probably arboreal in its habits',[56] Wallace appealed to supernatural mechanisms to explain humanity's unique mental attributes. In some ways, Wallace was too sophisticated for his own good. At a time when even the liberal Darwin could view the native peoples of the countries he visited as 'degraded savages', Wallace argued that all human beings are essentially equal.[57] Indeed, he counterposed the morality of the 'primitive' people he encountered to the 'social barbarism' of Victorian England, and their harmonious coexistence with nature to the environmental destruction being wreaked by the Industrial Revolution.

This positive view of human potential led Wallace to wonder why, if the complexities of the human mind were a product of blind chance, people living in primitive settings had the same mental capabilities as those in the civilized world. To him, this implied that the great part of human intelligence in such an environment went unused. To explain this conundrum, Wallace concluded that 'some higher intelligence directed the process by which the human race was developed',[58] much to the dismay of Darwin, who told Wallace, 'I hope you have not murdered too completely your own and my child!'[59] So it was that Darwin, the bourgeois gentleman, proved more revolutionary than Wallace, the socialist.

Despite its success, the theory of natural selection as expounded by either Darwin or Wallace faced a major problem: the lack of a proper explanation for how new characteristics are passed down to offspring so that those more appropriate for survival in a new environment come to predominate. At this time, human offspring, like animals or plants, were known to share many characteristics with their parents, whether in looks, abilities, or temperament, but this was thought to occur through a mixing of the parents' blood, an idea espoused by ancient Greek philosopher Aristotle,[60] and which underlies the still-used phrase 'blood relations'. Aristotle's view that offspring inherit characteristics from their parents by their mother's menstrual blood combining with their father's semen (which he considered a purified form of blood), may seem ridiculous now but it fitted the observations he could make with the limited experimental tools available.[60] However, the idea that inheritance was passed down in this way posed problems for evolutionary theory, for such mixing would dilute any new characteristics that arose. This problem plagued Darwin until his death.

Yet ironically, the problem of inheritance had, unknown to Darwin, been solved in his lifetime by Gregor Mendel.[61] Mendel's experiments on inheritance in pea plants at St Thomas's Abbey, Brno, now in the Czech Republic, first showed that characteristics like height, colour, and shape are passed down to offspring according to precise mathematical rules, and his work helped complete the puzzle that had baffled Darwin. Of crucial importance was Mendel's conclusion that an organism's inherited characteristics are determined by discrete 'factors', later termed genes. This was hugely important for evolutionary theory, for such discrete elements could be passed down to offspring without their effect being diluted by mixing. Mendel's work implied that, for any particular characteristic, there are two copies of each genetic determinant, one inherited from the father, one from the mother (see Figure 1). In 'dominant' situations, only one copy of a gene variant is needed to determine the characteristic, whilst in 'recessive' situations, both copies are required. While Mendel only studied peas, the mathematical rules he established explain the inheritance of human disorders like cystic fibrosis or Huntington's disorder, these being recessive and dominant respectively, so that we still use these rules to predict the likelihood of someone inheriting them.[62]

Mendel realized the potential general significance of his findings, saying 'I am convinced that it will not be long before the whole world acknowledges the results of my work.'[61] Instead, despite being published less than a decade after *The Origin of Species*, the importance of Mendel's findings lay unrecognized until they were rediscovered in 1900, Mendel having died in 1884. Quite why their value for evolutionary theory was not recognized sooner remains a matter of debate. Some have pointed to the fact the findings were published in the local

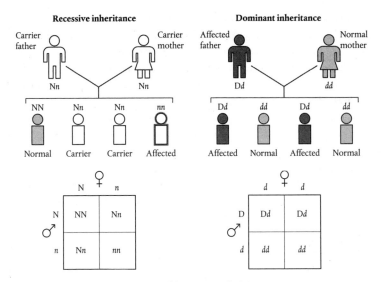

Figure 1. Mendel's patterns of inheritance

journal of the Brno Natural History Society; however, this journal was sent out to libraries across Europe and Mendel himself tried to interest leading international botanists in his work, albeit without success.[61] Another explanation is that the mathematical rigour Mendel applied to inheritance was too advanced for its time, biologists in those days being used to qualitative, not quantitative, explanations of nature.[63] Whatever the exact reason for their initial obscurity, the value of Mendel's findings was finally recognized more than thirty years after they first appeared in print, by Hugo de Vries, Carl Correns, and Erich von Tschermak, scientists coming to similar conclusions themselves.[64] It would be nice to say Mendel's prior claim to fame was acknowledged fairly and ungrudgingly, however, de Vries originally seems to have tried to pass off Mendel's insights as his own, only to be rumbled by Correns.[64] Such are the temptations of immortality, albeit of the scientific variety.

Given that Wallace lived for thirteen years after the rediscovery of Mendel's findings, one might have expected him to welcome this major theoretical link that had eluded him and Darwin. Yet, up to his death, Wallace remained unconvinced of its importance.[65] His scepticism was partly due to the fact that de Vries linked

Mendelism to his own view that changes in species occurred in leaps, which he explicitly counterposed to the gradualism that both Darwin and Wallace espoused. In addition, Wallace viewed the rigidity of Mendel's 'laws' as antagonistic to the plasticity he saw as central to evolution, noting that Mendel's factors 'are transmitted without variation, and therefore, except by the rarest of accidents, can never become adapted to ever varying conditions'.[65] It was a good point but neglected the crucial role of mutations: changes in genes that affect their function.

A proper recognition of the role of mutations in evolution awaited the work of Thomas Morgan, who, early in the twentieth century at Columbia University, New York, systematically identified and characterized these genetic changes in what became a central 'model' organism for genetics—the fruitfly.[66] Morgan was initially sceptical about both Darwinism and Mendelism, being more influenced by de Vries's theory that evolution was driven by dramatic changes affecting the whole body plan of the organism far more rapidly than could be accounted for by natural selection.[67] This idea was based upon de Vries's discoveries of new forms of the primrose plant. In fact, we now know these were due to genome duplications in these plants, with little general relevance for normal evolution. However, they inspired Morgan to go looking for mutations in animals. He first studied rodents, but then, realizing they reproduced far too slowly for identification of the molecules of inheritance, Morgan turned to fruitflies, because their short generation span, and the fact they can be kept in a milk bottle with some banana to keep them happy, meant they were ideal for scientific studies.[68] From 1907 onwards, he and his co-workers spent many hours searching for naturally occurring mutations.[69] At first it seemed a fruitless task and one designed to drive the team bananas, leading Morgan to exclaim, 'There's two years' work wasted. I have been breeding those flies for all that time and have got nothing out of it!'[66]

However, in April 1910 the team made a breakthrough. In one of the bottles was a fly with white eyes rather than the normal red ones.[68] Further analysis showed that this was recessive, while red eyes were dominant, in line with Mendel's laws. There was a complication though, for this particular characteristic showed an unusual pattern not identified by Mendel, being generally only present in males.[69] So what could be the material basis for this difference? The answer eventually came from one of those occurrences in science when two lines of investigation fortuitously converge. Although Mendel's findings were only rediscovered in 1900, in the intervening period major discoveries were still being made relating to the material basis of inheritance. By this time it was recognized that cells divided to form other cells, and that hereditary information was stored in their central nucleus. But that still left unanswered how that information was distributed during cell division to each daughter cell formed during this process. The answer came from Walther Flemming of Kiel University, Germany, in 1878. In the late nineteenth century Germany led the world in the production of new chemical dyes for the textile industry, and some scientists realized these dyes could also be used to stain cellular structures. Flemming used this method to study cell division.[70] By using dyes to both stain the cell's components and 'fix' the cell at a particular stage in the process, Flemming identified tiny condensed threads that formed in the nucleus during a stage of cell division that he named mitosis, after the Greek for thread. The threads were named chromosomes, Greek for 'coloured bodies', reflecting their discovery with the aid of dyes.

Flemming demonstrated that chromosome pairs are first duplicated during cell division, the pairs then subsequently segregating into the two daughter cells.[70] The obvious analogy with inheritance convinced many scientists that chromosomes must play a crucial role in this process. This was demonstrated experimentally in 1889 by Theodor Boveri, working at the Stazione Zoologica in Naples, who showed that chromosomes are required for embryo development to occur.[71] And in 1905, just before Morgan's discovery of sex-linked mutations, Nettie Stevens and Edmund Wilson showed that cells of male and female flies could be distinguished by females having two copies of the X chromosome, while males have one X and one Y chromosome.[72] Despite previously opposing the idea of chromosomes being agents of inheritance, based on his belief that passive observations were no substitute for direct experimentation, Morgan realized his findings must mean the gene for the red eye colour resided on the X chromosome, white eyes being an 'X-linked' recessive disorder.

We now recognize this pattern of inheritance in human disorders like haemophilia, Duchenne muscular dystrophy, and red–green colour blindness. While standard recessive disorders require both copies of a gene to be faulty since males only have one X chromosome, in X-linked disorders a single faulty copy can cause the disease. This is why it is very rare to find women with such disorders; instead, human females 'carry' these diseases to the next generation of males, as Queen Victoria did when she passed on haemophilia to many royal males in Europe.[73] In the fly experiments, however, a few white-eyed females were detected, and in such proportions that confirmed they must have two faulty copies of the gene associated with red eyes. So, despite his initial scepticism about the link between chromosomes and inheritance, Morgan had himself provided the first experimental evidence for this link.[74] And indeed, in a rare moment of generalization on his part, this extreme empiricist now proposed that all genes must reside on chromosomes. Finally a connection had been made between Mendel's abstract 'factors' and real cellular entities within the cell.

A further major step forward came with the discovery by Morgan's colleague, Hermann Muller, that treating flies with X-rays greatly increased the chance of their offspring becoming mutants.[75] This was the first demonstration of the dangers of radiation, and Muller would build a career highlighting these dangers for human beings. But it also allowed Morgan's lab to dramatically increase the numbers of available mutants, and to create flies with multiple mutations and then compare their inheritance.[76] This led to some unexpected findings. Mendel's laws predicted that inheritance of each characteristic should be completely independent from the next,[61] but that those on the same chromosome should be inherited together, and this was found to be the case in most of the fly studies.

Puzzlingly though, while some mutations on the same chromosome stayed 'linked' in this way, others became separated, and the degree to which this happened varied. Seeking an explanation for these strange findings, Morgan was intrigued to learn of an observation made in 1909 by Frans Janssens at Leuven University, Belgium; he had been studying the segregation of chromosomes during the formation of eggs and sperm.[77] Normal cell division—mitosis— involves, first, the duplication of a cell's genetic material, and then equal segregation of this to each daughter cell, which are therefore exact copies of the parent cell (see Figure 2). But eggs and sperm only contain half the genetic material of a normal cell, because there is no duplication, only segregation. However, what Janssens found when studying this process, named 'meiosis' from the Greek for lessening, was that, before splitting, each chromosome pair became twisted together and even appeared to exchange segments with each other.[77] Morgan realized that this 'crossing over' explained why some characteristics were more linked than others. If the genes on a chromosome are like beads on a string, the closer two genes are together, the more likely the chance of remaining together during crossing over.

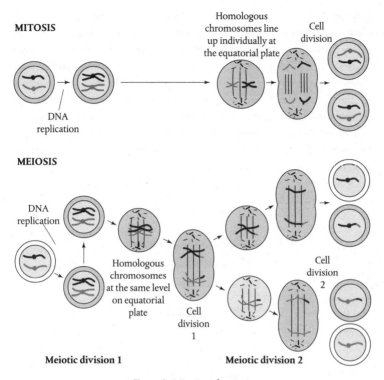

Figure 2. Mitosis and meiosis

This insight led to the first genetic 'map'. Remarkably, Alfred Sturtevant, who first proposed such a map, was only an undergraduate student when he came up with this idea.[68] Sturtevant had been so inspired by Morgan's lectures he asked if he could work in the latter's lab. Morgan was happy to take gifted students under his wing and Sturtevant became fully integrated into the lab even as he studied for his degree. As further inspiration for an ambitious student, Morgan's lab was one of the first in which students were treated as colleagues and encouraged to be co-authors or even sole authors of papers.[78] Such democracy was unusual given that the model at this time was the 'German research university, in which the Geheimrat, the great scientific leader, ordered the hierarchy of his subordinates'.[68] Instead, in Morgan's lab, 'each carried on his own experiments, but each knew exactly what the others were doing, and each new result was freely discussed'.[79]

Musing one evening about the linkage phenomenon, Sturtevant realized that 'the variations in the strength of linkage already attributed by Morgan to difference in the spatial separation of the gene offered the possibility of determining sequence in the linear dimensions of a chromosome. I went home and spent most of the night (to the neglect of my undergraduate homework) in producing the first chromosome map.'[68] Crucially, knowing the linkage frequencies for at least three characteristics would allow not only the order of the genes associated with these characteristics on the chromosome to be known, but also their relative distances from each other (see Figure 3).[78] As well as revolutionizing experimental genetics, this discovery paved the way for genetically mapping human characteristics, including diseases. The pioneering nature of Morgan's research was acknowledged in 1933 with a Nobel Prize; notably, he shared the prize money

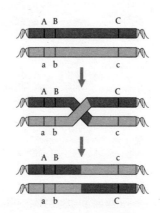

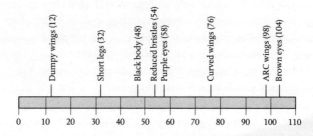

Figure 3. Crossing over and genetic map

with Sturtevant and another student, Calvin Bridges.[68] Muller also received a Nobel Prize for his discovery of the mutagenic properties of radiation, in 1946.[75] Only a year before this, in 1945, the atomic bombs dropped in Japan had demonstrated the devastating effects of radiation on the human body.

The first human disease shown to follow a Mendelian pattern of inheritance was alkaptonuria, a defect in metabolism associated with early-onset arthritis and heart disease, but whose most visible symptom is a tendency to produce black urine. Its inheritance pattern was first identified in 1902 by Archibald Garrod, a physician whose bedside manner was said to be limited to interest in his patient's urine samples.[80] But while he might have lacked interpersonal skills, Garrod played a key role in advancing the scientific basis of medicine. For him, science was 'a way of searching out by observation, trial and classification; whether the phenomena investigated be the outcome of human activities, or of the more direct workings of nature's laws', and he showed the power of this approach by recognizing, just two years after the rediscovery of Mendel's work, that alkaptonuria displayed the very same recessive pattern of inheritance shown by the latter's pea plants.[80]

In fact, Garrod's insights were too advanced for most biologists and clinicians at that time; they saw his findings as relevant for this one odd disorder, but not generally applicable to human disease.[80] Instead, it was only in the 1920s, when Ronald Fisher, Sewall Wright, and J. B. S. Haldane began independently seeking mathematical explanations for the inheritance patterns in human populations by reference to the laws that Mendel had established, that the union between Darwinism and Mendelism could be considered secure.[81] Not everyone was enamoured of this approach, evolutionary biologist Ernst Mayr likening it to comparing 'the genetic contents of a population to a bag full of colored beans. Mutation was the exchange of one kind of bean for another . . . To consider genes as independent units is meaningless from the physiological as well as the evolutionary viewpoint.'[82] Haldane agreed that such 'beanbag genetics' were based on many unrealistic 'simplifying assumptions', namely that the characteristics being studied followed simple Mendelian rules, mating was random, and populations could be treated as effectively infinite; however, he also maintained that this approach could yield important quantitative information about the incidence and spread of disease, and, indeed, other characteristics in a human or animal population.[83]

Another major step forward in genetics was the demonstration by George Beadle and Edward Tatum of Stanford University that genes and enzymes were directly connected. It was while listening to a seminar by Tatum in 1941 in which the latter posed the question 'What do genes do?' while also discussing cellular biochemistry, that Beadle suddenly realized how to connect the two issues. Observing Tatum writing sequences of reactions on the blackboard, Beadle 'suddenly realized how stupid we had been all these years...instead of looking for reactions by enzymes controlled by known genes, why not look for genes that control already known chemical reactions? We might then expect to find mutations characterized by an inability to synthesize essential diffusible substances such as vitamins, amino acids and other building blocks of the cell's protoplasm.'[84] A former student of Morgan, Beadle persuaded Tatum that the best way to investigate this issue was to use an even simpler organism than the fly—a bread mould called *Neurospora*—whose biochemistry was well worked out. By using X-rays to create mutants in this species, then studying which biochemical pathways were defective, in only a few months Beadle and Tatum showed that mutants segregating in a Mendelian fashion affected specific enzymes in these pathways. This suggested each gene is used to produce a single protein, and resulted in a Nobel Prize for Beadle and Tatum in 1958.

What was remarkable about Mendel's pioneering work in establishing the principle of genes as discrete entities; the extension of this principle by Morgan, Beadle, and Tatum experimentally, and Garrod clinically; and its mathematical underpinning by Haldane and colleagues, was that all this had been achieved despite the molecular basis of genes remaining unknown. For geneticist Richard Goldschmidt, reflecting back on this era in 1950, it was such a leap forward in science as to rank alongside 'the explanation of the movements of the celestial bodies by Kepler, Copernicus, and Newton; Galileo's experiments inaugurating the age of inductive science, and Darwin's establishment of the theory of evolution'.[68] Yet as the 1950s began, and post-war austerity gave way to the post-war boom and the new music of rock 'n' roll, the molecular foundation of this new genetics remained far from clear, almost a century after Mendel's breakthrough. This state of affairs was, however, about to change in a way that would first strengthen the modern synthesis, but eventually come to challenge it, with the discovery that the genetic material was deoxyribonucleic acid, more popularly known as DNA.

2

LIFE AS A CODE

'Science moves with the spirit of an adventure characterized both by youthful arrogance and by the belief that the truth, once found, would be simple as well as pretty.'   *Jim Watson*

'The human genome consists of about 3.3 billion base pairs...0.8 gigabytes of information, or about what you can fit on a CD. With a microwave radio transmitter, you could beam that amount of information into space in a few minutes, and have it travel to anyone at light speed.'   *Seth Shostak*

We're so used to thinking of DNA as the blueprint of life that it's easy to forget how resistant many biologists were to this idea at first. DNA was discovered by Swiss scientist Friedrich Miescher as early as 1869, through his interest in identifying the biochemical components of the cell nucleus. Miescher made the discovery while studying with the biochemist Felix Hoppe-Seyler in Tübingen, Germany. Like something out of a Frankenstein film, Hoppe-Seyler's lab was based in a medieval former royal castle; Hoppe-Seyler occupied the former laundry room while Miescher worked in the old kitchen.[85] Hoppe-Seyler wanted to catalogue the chemicals in blood, and had already studied red blood cells, so Miescher was given the task of looking at white blood cells. This proved fortuitous since these cells have a very prominent nucleus, and it was while studying this that Miescher identified DNA as a central component and set out to purify it. Fortunately, or unfortunately depending on your point of view, a ready supply of white blood cells was available from pus in surgical bandages collected at a local hospital. Every day a hospital orderly delivered these gruesome items which Miescher smelled to determine which were the freshest. But 'driven by a demon', he threw himself into the task and following a year's hard work he had a sample of pure DNA.[85]

Miescher predicted that DNA would soon 'prove tantamount in importance to proteins'.[86] However, despite a major clue as to the physiological role of the molecule being its exclusive location on chromosomes, DNA was not thought complex enough to carry the hereditary information. Instead, proteins, also abundant in chromosomes for reasons we'll explore later, seemed much more suited to this role.[86] Proteins were, after all, at this time being identified as life's 'building blocks'. We now know proteins form the primary cellular structures, but also catalyse chemical reactions, and control transport into and within the cell. To carry out such multiple roles, proteins come in many shapes and sizes, a feature made possible by the 20 different amino acids, each with its own individual character, of which these molecules are built. The multiple ways in which amino acids can be combined produces the dizzying diversity of protein types (see Figure 4). Proteins can be long, thin, and fibrous like collagen, with a higher tensile strength than steel, which it imparts to bones or cartilage, or soluble and globular like haemoglobin, which carries oxygen around the body. It was hardly odd then to assume that since proteins are the building blocks of life, they would also be its instruction manual.

In contrast, DNA seemed a far simpler and less interesting molecule. Unlike proteins, the units of DNA come in just four varieties. A unit of DNA is called a nucleotide and consists of three parts. The first is a deoxyribose sugar, which

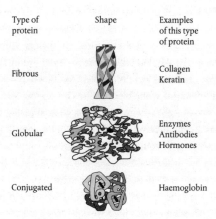

Type of protein	Shape	Examples of this type of protein
Fibrous		Collagen Keratin
Globular		Enzymes Antibodies Hormones
Conjugated		Haemoglobin

Figure 4. Different types of proteins

forms a pentagon, in contrast to the hexagonal glucose. The second is a phosphate, which joins the units to each other, such that a DNA chain is said to have a 'sugar phosphate backbone'. The third is a nitrogenous base; this part varies, being either adenine, guanine, cytosine, or thymine, although nowadays these are generally abbreviated simply to A, G, C, and T. While proteins vary in shape and size, DNA is a long, thin molecule that just goes on and on. Since the bases themselves didn't appear that different chemically, and Phoebus Levene, a chemist at the Rockefeller Institute in New York, had proposed that the sequence of bases was an endless repetition of ACGT, it wasn't at all clear how this apparently boring molecule could do anything significant at all.[87] So, as late as the mid-twentieth century, a commonly held view was that, despite being a central component of chromosomes, 'DNA could only be some sort of structural stiffening, the laundry cardboard in the shirt, the wooden stretcher behind the Rembrandt, since the genetic material would have to be protein.'[87] This scepticism was such that even when clear evidence emerged that DNA was the molecule of inheritance, it was disregarded by most biologists. However, nature has a habit of asserting itself despite the intentions of the scientists studying it. This was certainly true for Oswald Avery, also at the Rockefeller, who in 1944 began studying transmission of heredity in bacteria as a way of identifying the molecule responsible.

Previously, Fred Griffith of the British Ministry of Health had shown, in 1928, that bacteria could swop particular characteristics in the bacterial equivalent of sex, these being then transmitted to future generations. It was a discovery with major future implications for the study of genetics, but at the time its significance was barely recognized, partly because of Griffith's extremely shy nature.[88] On one occasion, his colleagues had to virtually kidnap him and drive him to a conference where he'd been invited to speak. Even once there, he mumbled his way through some obscure aspect of his studies, and made no mention of his key finding. Avery, however, did hear about the findings, and he reasoned that the process Griffith had identified could be used to study the molecular basis of heredity by eliminating each individual component of the bacteria, to see how this affected transmission of genetic information. Particularly important was the fact that transmission was possible even with dead bacteria. Avery thus destroyed, first, the sugar molecules coating the bacteria, then the proteins, then their RNA. None of this affected transmission, until finally the bacterial DNA was destroyed, upon which the ability to pass on genetic information ceased.[88]

Yet even this apparently clear-cut result failed to sway many biologists. It didn't help that Avery was ultra-cautious in his report of the study, concluding that 'it is of course possible that the biological activity ... is not an inherent property of the nucleic acid but is due to minute amounts of some other substance', something seized upon by those who believed that proteins were the agents of inheritance, and therefore must be the contaminating factor.[89] In addition, Avery's findings left unresolved how the apparently simple DNA could be life's blueprint. Max Delbrück of the California Institute of Technology observed later that Avery's findings clashed strongly with the belief that 'DNA was a stupid substance, a tetranucleotide which couldn't do anything specific'.[89]

One person unconvinced by Avery's findings and the claims for DNA's importance was Alfred Hershey, of Cold Spring Harbor Laboratory.[90] In 1952 he decided to further test the matter using a type of virus called bacteriophage, or phage for short. Phage infect bacteria, proving there is some truth in Jonathan Swift's claim that 'a flea has smaller fleas that on him prey; and these have smaller still to bite 'em, and so proceed *ad infinitum*'.[91] Studies of phage under the electron microscope by Thomas Anderson at Cold Spring Harbor had recently revealed that the phage never entered the bacterium during an infection; instead, it remained on the host cell's surface. This suggested that it acted like a hypodermic needle to inject the hereditary molecule into the bacterium; this then took over the cellular machinery in order to create new viruses. Hershey realized that, by labelling the viral DNA and protein with radioactive phosphorus and sulphur respectively, it should be possible to determine which had entered the bacterium.[90] One technical obstacle remained for Hershey and his assistant Martha Chase, and that was finding a way to separate the bacteria and phage after infection. Having tried 'various grinding arrangements, with results that weren't very encouraging', Hershey and Chase decided to use a simple kitchen blender. This device, more generally used for making Mai Tais, mojitos, and other cocktail party drinks of the 1950s, efficiently removed phage from the bacteria without rupturing the latter. With this tool, Hershey and Chase showed conclusively that the viral DNA, not its protein, entered the bacterium. Even faced with this apparently incontrovertible evidence, Hershey initially had trouble accepting the result, telling the audience to which he first presented the findings that 'I don't believe in that DNA'.[90] The problem was that it still remained unclear how such a boring molecule as DNA could act as the hereditary molecule.

Instead, it was only when Jim Watson and Francis Crick, working at Cambridge University, unveiled their model of the famous double helix structure of DNA, that it finally became clear how the molecule could function in such a way. Watson and Crick's discovery, published in *Nature* in April 1953, is rightly seen as the starting point of a revolution in how humans view life and their own species. Thanks to Watson's own frank and somewhat scurrilous account in his book *The Double Helix*, this discovery has become famous for showing that brilliant insight and dubious practices can triumph over the sort of plodding, painstaking construction of theories, based upon a careful examination of experimental data, often held up as an example of the scientific process.[92] A precocious child, Watson began studying biology at the University of Chicago, his native city, aged only 15. There he was seduced by genetics and applied to do a PhD at Indiana University in 1947 because of its association with Hermann Muller.[93] But already the study of fruitfly genetics was looking jaded in comparison to new directions, such as the use of bacteriophage. Consequently, Watson began his PhD studies with Salvador Luria and Max Delbrück, pre-war refugees from fascist regimes in Italy and Germany respectively, who were now establishing themselves as masters of this new field of research.[93] Only two years into studying for his PhD, Watson decided that 'I wanted to find the structure of DNA; that is, DNA was going to be my objective.'[94] Comparing the quest to the fever that had gripped his nation a half century earlier and led thousands to brave all in the Yukon region in Alaska, Watson claimed 'DNA was my gold rush'.[95]

Realizing that the best place to achieve his objective was in England, where structural analysis of biological molecules was, at that time, amongst the best in the world, Watson joined the Cavendish Laboratory in Cambridge, presided over by Sir Lawrence Bragg. Although, in Watson's book, Bragg is portrayed as a stuffy administrator, more likely to be at his London club than doing any actual science,[96] in fact the latter had already distinguished himself by solving the first ever molecular structure, that of NaCl or common salt, in 1912, and so became the youngest person to win a Nobel Prize at the age of 25.[97] The key to Bragg's success was a technique called X-ray crystallography, which works by firing a beam of X-rays at a crystal of the molecule under investigation; the scattering of this beam in response to the atoms it encounters is used to build up a picture of the structure of the molecule.[97] At the Cavendish this technique was primarily being used to solve the 3D structure of a protein, haemoglobin, while, due to an arrangement with the Medical Research Council, analysis of DNA was being carried out at the Randall Institute, King's College London, by Maurice Wilkins and Rosalind Franklin.[98] However, such minor details would soon prove less important than Watson's thirst to get to the goal first. It also helped that he found an ideal partner at the Cavendish in Francis Crick. At this point Crick appeared to be the opposite of precocious, since, although 35 years old, he hadn't even completed a PhD. Yet over the next decade Crick would prove to have one of the sharpest minds in twentieth-century science.

An important factor in Watson and Crick's favour was that the research effort at the Randall Institute was seriously undermined by the inability of Wilkins and Franklin to work together. The seeds of this schism were sown by John Randall himself, one of the creators of radar, who after the war was given funds to use physics to solve key questions in biology.[99] Randall led both Wilkins and Franklin to believe they would be leading the effort to determine DNA's structure, and, coupled with a clash of two quite different personalities, this led to a rapid falling-out.[99] One of the greatest travesties in the discovery of the DNA double helix was that Watson and Crick obtained a crucial piece of unpublished experimental data from Rosalind Franklin without her knowledge.[100] That this data was shown to Watson by Wilkins, unbeknown to Franklin, is now recognized as one of science history's more glaring injustices. Wilkins has his own version of events, which stresses Franklin's prickly character,[101] but it is hard not to view the way Franklin was marginalized at King's as linked to her gender. So she was barred from lunching with her male colleagues,[102] and the nicknames many called her— 'Rosy', the 'Dark Lady'—show that it was not just Watson at this time who treated her in a sexist manner. Yet these features of the discovery should not detract from the brilliance of Watson and Crick's insight when their model-building strategy, which had been largely dismissed by Franklin and Wilkins, led them to the double helix structure.[103] In this respect the pair were undoubtedly helped by DNA having 'molecule of life' written all over it, so much so that, on Saturday, 28 February 1953, when they finally resolved the structure, Crick told everyone within earshot, at their subsequent celebration at the Eagle pub in Cambridge, that he and Watson had discovered the 'secret of life'.[104]

The first major revelation of the double helix structure was showing the way in which the molecule is capable of self-replication. We've seen that nucleotides come in four varieties, defined by the bases adenine, cytosine, guanine, and thymine, or A, C, G, T for short. Watson and Crick proposed that the two strands of the double helix were held together by an attraction of A for T, and G for C, these paired bases occupying the space within the helix-like steps in a spiral staircase (see Figure 5). So famous is this iconic structure now that Martin Kemp, an art historian at Oxford University, recently called it 'the Mona Lisa of modern science'.[105] Scientifically, the structure made sense of a previous discovery by Erwin Chargaff of Columbia University, New York, whose chemical analysis of DNA in 1952 showed that A and T occurred in equal amounts, as did G and C.[106] Chargaff himself thought Watson and Crick were a couple of cowboys on the make, after an encounter in which the pair showed their ignorance about the chemical structures of the nucleotide bases.[107] A decade after Watson and Crick's discovery, his opinion of them hadn't improved, as shown by his comment that 'molecular biology is essentially the practice of biochemistry without a license'.[108] Yet their structure made sense of Chargaff's findings in a way that had eluded him.

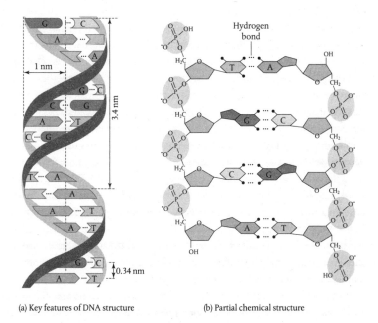

(a) Key features of DNA structure   (b) Partial chemical structure

Figure 5. Structure of DNA showing double helix and base pairing

That AT and GC 'base pairing' might be centrally linked to how DNA replicated itself, was suggested by one of the most famous cryptic comments in science, when, in Watson and Crick's *Nature* paper, they said 'it has not escaped our notice that the specific pairing we have postulated immediately suggests a possible copying mechanism for the genetic material'.[109] Indeed, their follow-up paper, also published in *Nature*, proposed a form of replication they named 'semi-conservative', that involved the two strands splitting apart during each cell division, and forming the template for another mirror-image strand to be formed from each.[110] Another less likely possibility was that the double helix did not split into two during replication but instead another new molecule formed alongside it—'conservative' replication.

One person for whom Watson and Crick's discovery had life-changing import-ance was an Oxford University PhD student, Sydney Brenner, who would later play a major role in solving the genetic code and many other fundamental problems of biology. He recalled how having been told that the structure of DNA had probably been solved by two people in Cambridge, Francis Crick and Jim Watson, 'I went to Cambridge and saw the model and met Francis and Jim. It was the most exciting day of my life. The double helix was a revelatory experience; for me, everything fell into place and my future scientific life was decided there and then.'[111] Not everyone was as convinced though, for, as Brenner also noted, 'when the paper appeared a few weeks later, it was not well received by the establishment, composed largely of professional biochemists. They could not see, at the time, how profoundly it would change their subject by offering us a framework for studying the chemistry of biological information.' This may account for the delay in Watson, Crick, and Wilkins being awarded a Nobel Prize for the discovery, which finally happened in 1962. Whether Rosalind Franklin should also share the prize was never posed, due to her death in 1958 aged only 38, from cancer possibly induced by the X-rays she had been exposed to in her work.[112]

For the scientists who grasped the importance of the double helix discovery, there was much work to be done. Brenner later gave a sense of the excitement among those involved, when he noted that 'many have gone on to do important scientific work but all remember those wonderful times when we and our science were young and our excitement in meeting new challenges knew no bounds'.[113] A first important task was to confirm experimentally whether DNA replication occurred semi-conservatively or by the alternative conservative route. This question was answered by a study in 1958 by Matthew Meselson and Frank Stahl, often called 'the most beautiful experiment in biology' for its elegant simplicity. Meselson and Stahl met in the summer of 1954 as visiting graduate students at the Marine Biological Laboratory at Woods Hole, Massachusetts, where Watson and Crick were both giving guest lectures. Inspired by the talks and fired up by discussions over numerous gin martinis, the two decided to tackle this key question with the latest experimental techniques.[114] Since DNA contains nitrogen, which naturally occurs in two 'isotopic' forms, ^{14}N and ^{15}N, Meselson and Stahl used a new technique called density gradient centrifugation to distinguish DNA molecules containing the heavy and light isotopes.[115] Bacteria were grown in a broth containing ^{15}N for several generations, then transferred to one containing ^{14}N. By removing bacteria at various points and analysing their DNA, Meselson and Stahl showed that, after a single replication, the bacterial DNA had a density halfway between the high and low forms, exactly as expected of semi-conservative but not of conservative replication; the latter of which would have led to equal amounts of DNA of the higher and lower densities, but none of intermediate status. In 1956 the enzyme that carries out DNA replication—DNA polymerase—was isolated and characterized by Arthur Kornberg of Washington University, Saint Louis, which led to him being awarded a Nobel Prize in 1959.[116]

Yet while the DNA structure immediately suggested its likely mechanism of replication, there seemed nothing inherent in it to show how its information was translated into the biochemical processes of a cell or organism. The key was surely proteins, the building blocks of the cell and organism. But how could a linear sequence of DNA nucleotides code for the multiple shapes and sizes of different proteins? A vital clue was supplied by Fred Sanger, also in Cambridge, who was working on a method to identify the sequence of amino acids in a protein. A trained chemist, Sanger devised ways to mark the amino acid at the end of a protein chain and also break the chain into overlapping fragments.[117] He applied this technique to insulin, one of the few pure proteins available, due to its use by diabetics. Prior to Sanger's investigations, it was known that the proportion of different amino acids in a particular protein was specific to that protein, but the order in which they were strung together was thought to be random. However, in work that led to him being awarded a Nobel Prize in 1958, Sanger showed that every insulin molecule had the same unvarying sequence.[118] This raised the exciting possibility that since DNA also had a linear sequence, there might be a connection between this and the sequence of amino acids. The question was how the two could be connected.

Surprisingly, the first step in solving this mystery came not from a biologist, but a theoretical physicist, George Gamow, a refugee from Stalin's Russia, now based at the George Washington University, Washington DC. Gamow had played a key role in developing the 'Big Bang' theory of the origin of the universe.[119] Commenting on the timescale of the universe Gamow observed that 'it took less than an hour to make the atoms, a few 100 million years to make the stars and planets, but five billion years to make man!'[120] Perhaps it was a wish to understand what happened in those several billion years that led to an interest in Watson and Crick's new structure and its relevance for life, for Gamow began sending the two scientists letters outlining how a DNA code might operate.[121]

Initially dismissing him as a crazy stalker, Watson and Crick quickly realized Gamow had important insights to share. Gamow suggested that overlapping triplets of DNA bases specified a single amino acid, with the DNA acting as a direct template for the growing protein chain.[119] His view was that on each base a cavity must exist complementary in shape to part of an amino acid, so providing a mechanism for how a linear chain of nucleotides could code for one of amino acids. But there were problems with this model from the start. One was that proteins were known to be made not in the cell's nucleus, where DNA is located, but in the surrounding cytoplasm.[122] In fact, even removing the nucleus from a cell had no immediate effect on the speed at which proteins were made. These facts were hard to square with DNA acting as a direct template for protein synthesis.

What seemed necessary, therefore, was an intermediary between DNA and proteins. An obvious candidate was DNA's chemical cousin, RNA, known to be present in both nucleus and cytoplasm. RNA differs from DNA in that its units are ribonucleotides not deoxyribonucleotides, and it contains uracil, usually abbreviated to U, instead of the thymine found in DNA. In addition, RNA usually occurs as a single strand, unlike double-stranded DNA. Finally, while DNA stretches for the length of a whole chromosome, RNA occurs as much smaller fragments. So could RNA be acting as a go-between, ferrying information from DNA to proteins? This idea was strengthened by the discovery of RNA polymerase, an enzyme that catalyses the production of RNA from DNA.[123] But it still remained unclear how a nucleotide code, albeit one contained within RNA, could be transformed into one based upon amino acids.

In principle, Gamow's idea of overlapping nucleotide triplets acting as the template for an amino acid chain could apply to RNA as much as to DNA. But there were other problems with the model emerging, since it predicted that many pairs of amino acids would never be found next to each other in proteins. Yet as more proteins were sequenced, it became clear that any combination of amino acids was possible.[124] It was also very difficult to imagine how RNA could act as the direct template for protein synthesis in the way Gamow had envisaged for DNA, with the shape of the individual bases directly determining the sequence of the growing protein chain. Instead, in 1955, Crick proposed a radically different model, whereby amino acids were ferried to the point of protein synthesis by 'adaptor' molecules, which he suggested could themselves be some type of RNA.[125]

Crick's proposal was subsequently confirmed by Paul Zamecnik and colleagues at Massachusetts General Hospital, Boston. Zamecnik was a clinician who became interested in how cells regulate growth, and why this process seemed defective in some of his patients.[126] This led him to try and identify all the cellular components required for generation of proteins. A major step forward came with his discovery that a 'cell-free' extract of rat liver could still generate proteins if supplied with amino acids.[126] Plying this system with radioactively labelled amino acids or RNA in order to identify their respective roles in the synthesis process, Zamecnik noticed that, 'strangely enough, the RNA fraction was labelled from the amino acid precursor. In spite of careful washing procedures, the amino acid remained tightly bound to the RNA.' Finding that the specifically bound RNA molecules were of low molecular weight, Zamecnik realized they must be the adaptors Crick had proposed, and he named them transfer RNA, or tRNA for short. Subsequently, he showed there were twenty types of tRNA, one for each amino acid. But there was more to come, for he also demonstrated that proteins were manufactured in a huge molecular structure he called the ribosome, a complex of both proteins and another type of RNA, ribosomal RNA or rRNA.[126]

An unresolved issue remained, however. If the ribosome was the structure upon which proteins were made, and tRNAs the molecules that ferried amino acids to it, this failed to explain how the DNA code in the nucleus was subsequently translated into a protein sequence in the cytoplasm. Some intermediary must presumably carry the code from the nucleus to the cytoplasmic ribosomes, but what was its molecular identity? Watson and Crick initially proposed that rRNA itself fulfilled this role, with a unique version of one of these molecules being specific for each different protein. However, analysis of rRNA revealed it to be uniform in its sequence and very stable, not the qualities expected of the predicted transient intermediary. Instead, it was only with the discovery of a third type of RNA, by Sydney Brenner and colleagues, that the pieces of the puzzle finally fell into place.

Brenner was from a poor Jewish family in Johannesburg, his parents having immigrated to South Africa from the Baltic States.[127] His father, a shoemaker, spoke English, Yiddish, Russian, Afrikaans, and Zulu, but was illiterate, and it was an elderly neighbour who taught Brenner to read fluently before the age of 4, using the newspapers that served as a tablecloth in her house. A child prodigy, Brenner enrolled at the University of the Witwatersrand at the age of 15 and published his first scientific paper at 18. As we've mentioned already, Brenner was a PhD student in Oxford when he heard about Watson and Crick's great discovery and went personally to view their model. As he noted later, 'the moment I saw the model and heard about the complementing base pairs I realized that it was the key to understanding all the problems in biology we had found intractable—it was the birth of molecular biology'.[128] In 1960, Brenner decided to try and directly investigate the nature of the RNA intermediary with Matthew Meselson, whom we've already mentioned, and French geneticist Francois Jacob, whom we'll mention again shortly.[129] Brenner was intrigued by a report he had read of a study published in 1956, which showed that, at the height of a bacteriophage infection, there was a transient increase in an RNA with the same proportion of bases as the viral DNA. He suddenly realized that the transient RNA intermediary might reveal itself much more visibly during a bacteriophage infection because of the way the virus took over the workings of the cell.

To test this idea, Brenner, Meselson, and Jacob, at Meselson's laboratory at the California Institute of Technology, Pasadena, infected bacteria with the virus and, at the same time, added radioactively labelled phosphate to the media that bathed them.[130] The idea was to use the same density centrifugation method Meselson and Stahl had used to study DNA replication to identify the elusive messenger. But the ribosomes kept falling apart in the centrifuge, and for weeks it proved impossible to get things to work as planned. The frustration was, however, tempered by frequent trips to the beach, during which play was combined with intense discussions about how to overcome the block.[129] Finally, after trying a variety of conditions, the three experimenters detected a newly formed radioactive RNA associated with ribosomes, indicating it was linked to protein production. The intermediary had been found and was christened messenger RNA, or mRNA.[129]

These combined studies showed mRNA is produced in the nucleus as a single-stranded copy of the DNA code, with a distinct mRNA for each protein-coding gene. The tRNAs play a dual role, on the one hand bringing a specific amino acid to the ribosome, on the other recognizing the sequence of bases that specifies that amino acid in the RNA (and DNA) code (see Figure 6). The puzzle was almost complete; it only required the code to be cracked. Gamow had suggested a triplet code but did this match reality? In 1961, an ingenious approach was devised by Crick and Brenner to test this idea. Using chemicals to mutate DNA, they found they could insert or remove bases in phage DNA.[131] If they removed or added one or two bases the effects upon the virus were catastrophic; however, if three were

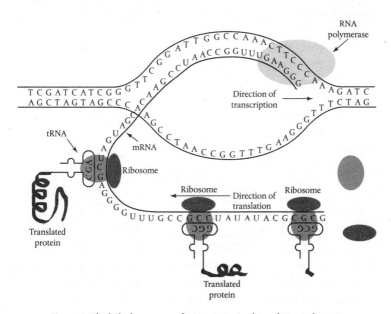

Figure 6. The linked processes of transcription and translation in bacteria

removed or added, there was little effect. This suggested each amino acid was coded by a triplet of bases, since adding or removing one or two bases caused a 'frameshift' in a three-letter word code, scrambling the meaning of the code beyond the mutation. In contrast, adding or removing three bases only altered a single amino acid. This still left the crucial issue of which triplets coded for which amino acid. This problem was solved by Marshall Nirenberg and his assistant Johann Matthaei at the National Institutes of Health in Maryland, who showed that an artificial RNA consisting of multiple U bases generated a protein consisting solely of phenylalanine, implying that UUU specified this amino acid.[132] Similar studies identified every other triplet sequences, or codons, that specify the twenty different amino acids, as well as where the protein starts and stops (see Figure 7). Nirenberg received a Nobel Prize for his discovery in 1973.

These findings showed Gamow was right to propose a triplet code but wrong to suggest it overlapped. Moreover, each amino acid could be specified by more than one triplet, making it a redundant code. The discovery of the genetic code signalled the primacy of the new discipline of molecular biology. Central to this was Crick's claim that life is a one-way flow of information from DNA to RNA to

Second letter

First letter		U	C	A	G		Third letter
	U	UUU ⎫ Phe UUC ⎭ UUA ⎫ Leu UUG ⎭	UCU ⎫ UCC ⎪ Ser UCA ⎪ UCG ⎭	UAU ⎫ Tyr UAC ⎭ **UAA Stop** **UAG Stop**	UGU ⎫ Cys UGC ⎭ **UGA Stop** UGG Trp	U C A G	
	C	CUU ⎫ CUC ⎪ Leu CUA ⎪ CUG ⎭	CCU ⎫ CCC ⎪ Pro CCA ⎪ CCG ⎭	CAU ⎫ His CAC ⎭ CAA ⎫ Gln CAG ⎭	CGU ⎫ CGC ⎪ Arg CGA ⎪ CGG ⎭	U C A G	
	A	AUU ⎫ AUC ⎪ Ile AUA ⎭ **AUG Met**	ACU ⎫ ACC ⎪ Thr ACA ⎪ ACG ⎭	AAU ⎫ Asn AAC ⎭ AAA ⎫ Lys AAG ⎭	AGU ⎫ Ser AGC ⎭ AGA ⎫ Arg AGG ⎭	U C A G	
	G	GUU ⎫ GUC ⎪ Val GUA ⎪ GUG ⎭	GCU ⎫ GCC ⎪ Ala GCA ⎪ GCG ⎭	GAU ⎫ Asp GAC ⎭ GAA ⎫ Glu GAG ⎭	GGU ⎫ GGC ⎪ Gly GGA ⎪ GGG ⎭	U C A G	

Figure 7. The triplet genetic code

protein—the 'central dogma of molecular biology'. He later claimed not to have realized 'dogma' meant a belief that cannot be doubted, and to have really meant a 'grand hypothesis'.[133] In fact, in his future theorizing Crick often proved far from dogmatic, but rather someone who, when solving scientific problems, learned 'how not to be confused by the details and that is a sort of boldness, and how to make oversimplifying hypotheses—and how to test them, and how to discard them without getting too enamoured of them'.[134]

There was undoubtedly a boldness in Crick's claim for the primacy of DNA, since the central dogma was centrally aimed at the biochemists who had dominated cell biology for over a hundred years. Such scientists saw the cell as a network of interacting chemical reactions, 'a subtle flux of materials and energy', all of it regulated by those ubiquitous proteins. However, they failed to explain how this related to the genetic material; therefore, what better way to distinguish molecular biology from biochemistry, than by asserting that however varied and complicated the actions of proteins are within the cell, ultimately they are merely slaves of the DNA code? Watson and Crick, and the scientists who flocked to their banner after the discovery of the double helix, saw this as a necessary step to undermine the old order that, in their view, blocked the path to a proper understanding of how life works. And in many respects they were right. However, from the outset there were some fundamental flaws in this new way of looking at life.

The view of life as the unravelling of a digital code has been expressed most forcibly by Sydney Brenner: 'If you say to me here is a hand, here is an eye, how do you make a hand or an eye, then I must say that it is necessary to know the programme; to know it in machine language which is molecular language; to know it so that one can tell a computer to generate a set of procedures for growing a hand, or an eye.'[135] Yet how true is the proposal that information only flows in a one-way direction from DNA to RNA to proteins? A lot depends on the definition of information. The proposal that DNA fulfils its role as the hereditary molecule by acting as a linear code, now led to it being compared to the blueprint of a building or machine. But even a blueprint must be read by an architect or engineer, so what was the equivalent for DNA? In fact, we now recognize that the cellular environment in which the DNA resides is crucial to unlocking its potential; and just how important this environment is, was shown by a startling discovery made by John Gurdon at Oxford University in 1962, just as the genetic code was being deciphered.

Gurdon serves as an example that future scientific success cannot always be gauged from school results, since he came last in biology out of the 250 boys in his year group at Eton College. As such, his teacher's assessment that 'he has ideas about becoming a scientist; on his present showing this is quite ridiculous' was harsh but at least appeared to have a factual basis.[136] Condemned this way, Gurdon reluctantly applied to do Classics at Oxford; however, at the last minute he was allowed to change to zoology, although only because a mistake had been made filling places in this subject and additional students were needed to fill the gap. Yet despite his teacher's criticisms Gurdon displayed an aptitude for experiments that led to him being offered a PhD project by Oxford biologist Michael Fischberg.[137]

Gurdon's somewhat ambitious project was to explore how DNA information in the nucleus of a fertilized egg can give rise to the multitude of distinctive cell types in an adult, this process being known as differentiation. As early as 1893, the German scientist August Weismann, most famously known for his distinction between the sex cells—sperm and eggs—and the remaining 'somatic' cells that populate the different organs and tissues of the body, had proposed that as embryonic cells differentiate into specialized cell types like nerve, muscle, heart, or liver, their hereditary material is progressively cast off or permanently inactivated, so that they become incapable of specifying anything other than that particular cell type.[138] Such was the dogma, but Gurdon decided to test whether differentiation was really irreversible by taking a nucleus from a differentiated frog cell, and seeing what happened when this was transplanted into an egg with its own nucleus removed. In fact, Robert Briggs and Thomas King of the Institute for Cancer Research, Philadelphia, had already shown that a nucleus from an early frog embryo—a blastula—triggered development when transplanted into an egg.[137] However, they had not tested the potential of nuclei from later stages of development when Gurdon began his experiments. Remarkably, Gurdon found that even nuclei from fully differentiated frog cells were capable of beginning embryo development anew, and generating a whole new fertile male or female frog.

This finding went so much against accepted dogma that, according to Gurdon, 'it took nearly 10 years for the major result to be accepted'.[139] Moreover, in an extension of their earlier experiments, Briggs and King subsequently failed to achieve successful development with nuclei from any later stage of development than the blastula. As Gurdon himself later acknowledged, it was therefore 'entirely reasonable for the sceptics to say, well these well-established people have already done this experiment and here's a graduate student from Europe who is disagreeing with them, why should we pay attention to that?'[139] Acceptance finally came, however, and Gurdon's demonstration that the cloning of adult animals was possible would eventually culminate in the cloning of Dolly the sheep by Ian Wilmut and Keith Campbell of the Roslin Institute, Edinburgh. This discovery would have to wait another thirty years due to greater difficulties in 'reprogramming' the nucleus of a differentiated mammalian cell compared to that of a frog, but this, and the demonstration by Shinya Yamanaka of Kyoto University that ordinary skin cells could be reprogrammed simply by changing the cellular environment, led to the eventual belated award of a Nobel Prize to Gurdon, together with Yamanaka, in 2012. Importantly, Gurdon's findings had shown, for the first time, that the passive potential of a genome to hold information can be distinguished from the active ability of that information to code for life's processes. And how this activation of the genome's potential is accomplished is the question to which we will turn in Chapter 3.

3

SWITCHES AND SIGNALS

'Scientific advances often come from uncovering a hitherto unseen aspect of things as a result, not so much of using new instruments, but rather of looking at objects from a different angle.'   *Francois Jacob*

'In science, self-satisfaction is death. It is the restlessness, anxiety, dissatisfaction, agony of mind that nourishes science.'   *Jacques Monod*

In science, as in life, sometimes it's the little things that count. Whether it be the apocryphal apple that landed on Newton's head and started him thinking about the laws of gravity, or the bread mould that blocked bacterial growth and led to Fleming's discovery of penicillin, apparently simple starting points can lead to the most profound scientific conclusions. One phenomenon initially deemed uninteresting but which turned out to be key to how the potential of the genome is unlocked, is how bacteria grow in a broth containing two different sugars— glucose and lactose. Rather than growing and dividing in a continuous manner, the bacteria grow rapidly initially but then their growth stalls briefly, after which they embark upon another burst of growth and division before this too comes to an end. This feature of bacterial growth was first noticed by Jacques Monod, a PhD student at the Sorbonne University in Paris, in 1941.[140] Monod was a late starter to a scientific career, being one of those infuriating individuals who excel in everything they do to such an extent that they find it difficult to make a choice. So his skills at music, as well as science, had his family seriously debating whether 'Jacques is going to be a new Pasteur, or a new Beethoven?'[141] Having finally decided to follow the scientific route, Monod's quest to explain the bacterial phenomenon he called 'diauxy', from the Greek for two growth phases, would occupy the rest of his career and eventually lead to a model of how genes are switched on or off that is broadly applicable to our own genomes. Yet, initially, Monod's studies were viewed as unimportant by other academics, with the head of his own laboratory confiding to one of his PhD examiners that 'Monod's work is of no interest to the Sorbonne'.[142]

However, Monod was not a person easily deterred by adverse circumstances. In occupied France, one could accept the Nazi presence, or fight it. Monod chose the latter route, despite being heavily involved in searching for a scientific explanation for the biological phenomenon he had discovered. Joining the French Resistance in 1942, a year later he became its Chief of Staff, the three previous occupiers of this post having disappeared without trace into the hands of the Gestapo.[143] Yet Monod did not suffer this fate—despite having to go underground at one point after individuals in his resistance 'cell' were captured—and he distinguished himself on multiple occasions in the fight against the Nazis. It was Monod who arranged parachute drops of weapons, railway and bridge bombings, and mail interceptions in preparation for the Allied invasion of France, and he also drafted the appeal to Parisian citizens to mount the barricades before the arrival of Allied forces into the city in 1944. Despite his military exploits, Monod somehow managed to continue with his studies; indeed, at times the two coincided, as when he hid vital resistance documents in the hollow leg bones of a giraffe skeleton outside his laboratory, this being one of the zoological specimens on display in the department.[144]

Still seeking to explain the two phases of bacterial growth he had observed, Monod speculated that this represented a switch from the metabolism of glucose to that of lactose. Somehow glucose must suppress the metabolism of the other sugar, but how remained a mystery. Monod's initial suggestion was that there must be some change in the conformations of the catalytic proteins—also known as enzymes—that carried out the metabolism of the two sugars. But another possibility presented itself in 1944 when he read an article in the travelling US army library which he had access to through his contacts with American soldiers, about Avery's discovery of the link between DNA and inheritance.[145] Could it be that the switch was due not to a change in the enzyme that metabolized lactose, but activation of the gene that coded for the enzyme, with the latter only being produced following such gene activation? If so, studying the switch might lead to important insights into how genes were switched on and off. Confirmation that the lactose metabolizing enzyme was being newly synthesized came when

Monod, now working at the Pasteur Institute in Paris, produced an antibody that detected the protein. This showed that the enzyme—now termed beta-galactosidase, or beta-gal for short—was only generated once all the glucose had been used up.[140] Somehow, this loss must send a signal to the cell that activated the gene coding for beta-gal. But could such a process be studied? Monod realized he needed to move from biochemistry to genetics. Luckily, he was now coming into contact with scientists who could help him make this transition.

In particular, a young Jewish scientist called Francois Jacob began working with Monod. Jacob had also played a heroic role fighting the Nazis, working first for the resistance and then participating in the D-Day landings.[146] Indeed, he was almost killed during the latter action after being hit by over a hundred pieces of shrapnel from a German air bomb; these permanently damaged his right side, including his hand, and put an end to his dream of becoming a surgeon. But medicine's loss was science's gain, for after the war Jacob began to forge a talented career as a geneticist. By the time he began working with Monod, he had already acquired great skill, not only in creating bacterial mutants but also in carrying out genetic crosses that previously had only been thought possible with multicellular organisms like fruitflies and mice. He did this by making use of the phenomenon of bacterial sex first identified by Griffith. In addition, Jacob proved an ideal partner in more than technical skills. According to Monod, Jacob was 'much more intuitive than I am; and I'm more of a strict logician than he is'.[147] In this sense, the combination of two quite different temperaments proved a potent mixture, just as had been the case for Watson and Crick, which shows that such partnerships can be as valuable in science as in music, with its Lennon and McCartney, or Jagger and Richards.

Isolation of mutants defective in the metabolism of lactose and the use of bacterial sex to carry out genetic crosses allowed Jacob and Monod to establish functional relationships between the different genes regulating the metabolism of this sugar. Importantly, they established a distinction between 'structural' genes—those coding for the enzymes that carried out the metabolism of lactose—and the 'regulatory' genes that coded for proteins which acted as switches to turn the structural genes on or off.[140] Another important discovery was that the beta-gal gene is switched on simultaneously with two others coding for proteins involved in lactose utilization. This co-regulation of proteins involved in the same metabolic process would turn out to be true of other bacterial genes. The unit of clustered lactose-metabolizing genes was christened the 'lac operon', and the term operon was soon used to define all clusters of co-regulated genes with a common function.[140]

A key moment in identifying how the lac operon was regulated was Jacob and Monod's isolation of a mutant whose lactose-metabolizing genes were always turned on, even in the absence of lactose. This suggested a defect in a protein that normally bound at the start of the lac operon and prevented it being turned on, or expressed. However, in the presence of lactose the protein lost its attachment and this allowed expression. This protein became known as the 'lac repressor', and established the idea that proteins could regulate the expression of genes.[140] Undoubtedly, the most unexpected aspect of the discovery was the demonstration that the interaction between the repressor and the gene it controlled was so direct. Jacob and Monod had assumed that the repression acted in some general fashion on the protein synthesis machinery. That this was not the case was truly exciting because, until then, as Monod put it, 'the gene was something in the minds of people—especially of my generation—which was as inaccessible, by definition, as the material of the galaxies'.[148] His and Jacob's demonstration that the gene was a tangible entity that could be turned on and off like a light switch was therefore a major revelation, and resulted in a Nobel Prize for both men in 1965.

This type of regulation was termed 'negative' because it involves the lifting of repression by an inhibitory protein that normally blocks gene expression. Excitingly, it also helped explain why, in certain cases of infection of bacteria by bacteriophages, the virus remained dormant, or 'lysogenic', until a stimulus, such as UV light, triggered expression of its genes. As Jacob put it, 'the analogy between [lactose repression] and immunity of lysogenic cells is such that we can hardly escape the assumption that immunity also corresponds to the presence of a repressor in the cytoplasm of lysogenic cells'.[149] Subsequent studies showed this was the case, with UV light triggering destruction of this repressor. Ironically, this phenomenon had been studied by André Lwoff, head of the Pasteur Institute, for some years.[150] So the two adjoining laboratories had essentially been studying the same molecular process without knowing it! Lwoff shared the Nobel Prize with Jacob and Monod for this discovery.

But was negative regulation sufficient to account for all instances of control of gene expression? Certainly Jacob and Monod thought so, but in the mid-1960s Ellis Englesberg and colleagues at the University of California, Santa Barbara studying an operon controlling metabolism of a different sugar, arabinose, presented evidence that positive regulation played an equally important role.[151] Despite this claim being based upon sound experimental data, it was initially dismissed by Monod, and for years Englesberg had trouble publishing his findings.[152] Yet Monod ought to have welcomed other forms of gene regulation, because one aspect of the phenomenon that had originally stimulated his interest remained a mystery. So although the discovery of the lac repressor explained why lactose stimulated expression of the lac operon, it left it unclear why, in a mixture of glucose and lactose, this expression only kicked off when all the glucose had been used up.[140] This dominance of glucose over lactose was named the 'glucose effect', but the model of gene regulation based solely upon the lac repressor provided no explanation for why such an effect should occur.

However, in 1965, a biochemist called Earl Sutherland showed that levels of a chemical called cyclic AMP, or cAMP, increase when glucose levels are low. We shall hear more about Sutherland and cAMP shortly, but for now it's enough to note that his finding excited the interest of two scientists studying the lac operon—Ira Pastan at the National Institutes of Health in Washington, and Agnes Ullmann, who worked at the Pasteur Institute alongside Jacob and Monod—who both immediately recognized it as a possible explanation for the glucose effect. Ullmann owed Monod a huge personal debt, for in 1960 he helped smuggle her out of her native Hungary where she was a dissident against the Stalinist regime and under threat of imprisonment after her role in the failed 1956 revolution.[153] Although Monod had been a member of the French Communist Party during the war, his later hatred of what he saw as the 'ideological terrorism' of Stalinism meant that he was willing to help. Using expertise learned in the French Resistance, Monod smuggled Ullmann and her husband across the tightly controlled Hungarian border and into Austria, hidden underneath a bathtub in a compartment of a pull-along camping trailer.[153] Now, however, she was about to challenge his view that negative regulation was the only mechanism controlling gene expression.

Independently, Pastan and Ullmann tested whether increasing cAMP levels artificially in bacteria in a mixture of glucose and lactose could activate the lac operon, and found that indeed it did.[154] This finding suggested that cAMP must work through an as yet unidentified regulatory protein. The search for this protein, by both Pastan and Jonathan Beckwith of Harvard University, eventually culminated in the discovery of what became known as the cAMP activator protein, or CAP.[154] Further studies showed that CAP was a positive regulator of exactly the type that Englesberg had proposed. Now, belatedly, the significance of his findings was recognized, and here, finally, was an explanation for Monod's original observation. In the presence of both sugars, initially only glucose is metabolized because its presence inhibits the activation of the CAP protein and therefore expression of the lac operon. But once the glucose is used up, the CAP protein is activated and the presence of lactose means the repressor does not inhibit expression of the lac operon (see Figure 8).[154]

So Monod's initial observation of a quirky feature of bacterial growth had led to the establishment of the fundamental principles by which gene expression in bacteria is regulated. Ironically, he was now offered the Chair of Biochemistry at the Sorbonne, the institution which had previously judged his studies as being 'of no interest'.[155] In fact, Jacob and Monod's discovery went even further than the recognition of how genes are turned on or off, for during his studies Monod stumbled upon another key biological process that would turn out to be central to

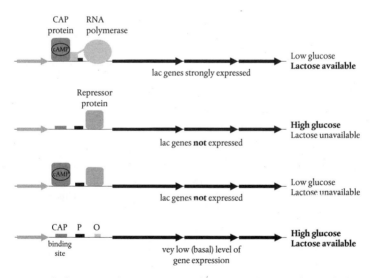

Figure 8. The lac operon showing negative and positive mechanisms of control of gene expression

how enzymes, and indeed proteins in general, respond to changes in their environment. What he had recognized was, that for lactose to exert its effects upon the lac repressor so rapidly, it must physically interact with the protein somehow. Moreover, cAMP must interact with the CAP protein in a similar fashion. Recognizing the importance of the discovery, Monod announced to a startled Agnes Ullmann that he had discovered 'the second secret of life'. Later Ullmann recalled that 'I was quite alarmed by this unexpected revelation and asked him if he needed a glass of whisky. After the second or maybe the third glass, he explained the discovery, which he had already given a name: "allostery".'[156]

Allosteric regulators act on enzymes at a site distinct from their catalytic centre. Instead, they influence their target's activity by altering its 3D structure, a 'conformational change' that alters the shape of the catalytic centre by action at a distance (see Figure 9). Just as allosteric regulation of metabolism is as important for our own cells as for bacteria, so the studies of gene regulation in bacteria pioneered by Jacob and Monod have proven highly relevant for complex

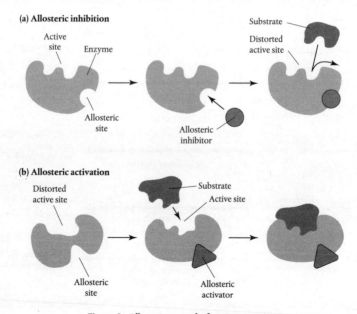

Figure 9. Allosteric control of enzyme activity

multicellular organisms, including our own species, albeit with some interesting twists. So cAMP also positively regulates gene expression in human cells; indeed, the connection between cAMP and the lac operon was merely further evidence of the importance of a signalling molecule originally discovered through a completely separate route, in mammalian liver.

We have already mentioned Earl Sutherland in passing, but his main contribution to science was the discovery of what have become known as 'second messengers', cAMP being just the first of these. Based at Western Reserve University in Cleveland, Ohio, Sutherland identified cAMP as the culmination of his quest to understand how the hormone adrenaline liberates glucose in the liver by stimulating the breakdown of glycogen, a polymerized form of this sugar molecule.[157] The importance of this carbohydrate store for everyday life is demonstrated by genetic diseases called glycogen storage disorders.[158] That sufferers will die unless fed a continual supply of glucose day and night shows how much we rely on our glycogen stores for normal existence. But until Sutherland's discovery, how signals outside the liver cell triggered changes inside it remained a mystery. He showed that cAMP was produced when adrenaline binds to proteins on the surface of the cell. Sutherland described the molecule as a second messenger by analogy with the hormone's role as the first messenger to the cell.[157] Initially, Sutherland's suggestion that a single molecule led to the numerous effects of adrenaline was met by disbelief. However, not only was he correct about the central role of cAMP in adrenaline's action, but subsequent studies showed that many other hormones stimulate its production. In addition, the idea of a second messenger turned out to have general relevance as other substances were shown to play similar roles in the cell. Sutherland was awarded a Nobel Prize for his discovery in 1971.[157]

We now know that many other substances can act as second messengers. These include other small molecules like cGMP, but also a charged atom or ion, calcium, and even a gas, nitric oxide. Indeed, the drug Viagra®, which became one of the best-selling drugs of all time because of its usefulness in treating impotence, works by stimulating the production of nitric oxide, which relaxes the blood vessels of the penis, allowing blood to flow into this organ, and thereby causing an erection.[159] Viagra® was originally developed as a treatment for angina—chest pains caused by restrictions in blood supply to the heart—but during initial clinical trials on healthy volunteers its ability to cause erections was noted, no doubt accompanied by reactions ranging from alarm to delight in those being tested. At first, there was scepticism about the potency of the drug among clinicians, which was famously overcome by Dr Giles Brindley at an American Urological Society meeting in Las Vegas in 1983, when he injected some of the drug just before he was due to give his talk, and 'over the course of his lecture demonstrated to his audience visible evidence that such an injection could induce an erection'.[160] Subsequently, Viagra® was developed in a form to be taken orally, greatly aiding its appeal as a drug not only for treating impotence but also for more recreational use. Ironically, there are now concerns that inappropriate overuse of the drug could itself lead to permanent impotence.

Second messengers act as a relay system passing on signals from the cell surface to target proteins inside the cell. A particularly important set of target proteins activated by second messengers are called kinases; these enzymes add a phosphate group to other proteins and thereby alter their properties.[161] Protein kinase A, or PKA, is activated by cAMP. We've already seen how, in bacteria, cAMP directly binds to the CAP protein to positively switch on the lac operon. A similar positive role is played in human cells by the cAMP regulatory element binding protein, or CREB.[162] However, rather than being directly activated by cAMP, CREB is instead phosphorylated by PKA after the latter has been activated by the second messenger (see Figure 10). This indirect method of control of gene expression has the advantage of being more finely tuneable than in bacteria. This extra level of complexity almost certainly reflects the more complex nature of the signals controlling cellular processes in multicellular organisms like ourselves, in which the cell not only responds to signals from the external environment but also from other cells within the body.

Just how complex a role gene regulatory proteins, also known as 'transcription factors', can play in multicellular organisms, was shown by recent studies that investigated the role of blood chemicals in the ageing process by the gruesome method of surgically attaching two mice together, one old, the other much younger, so they shared the same blood circulation system.[163] This approach was used in parallel with the more conventional method of giving blood transfusions from young to old mice. In both cases, the blood reversed age-related declines in memory and learning. After the treatment mice that were 18 months old, the equivalent of 70 years in a human, had acquired the enhanced learning abilities of a mouse that was only a few months old.[163] Saul Villeda of the

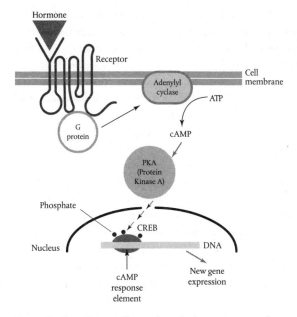

Figure 10. The cAMP signalling pathway leading to gene expression

University of California, who led the study, believes this shows 'there's something about young blood that can literally reverse the impairments you see in the older brain'.[164] Remarkably, these changes are primarily due to reactivation of CREB in a brain region called the hippocampus, that plays a central role in learning and memory, with the reactivated CREB turning on genes that regulate connections between nerve cells. This effect of young blood was traced to a protein called GDF11.[163] Before we get too excited about the possibilities for ageing humans, and whether Count Dracula was on to something after all (though presumably he would have needed to inject his victims' blood rather than drink it), it remains to be seen whether this kind of approach could be used in our own species. As Villeda said, 'I wish our manuscript could come with a big caption that says "Do not try this at home". We need a clinical trial to see if this applies to humans, and to see if there are effects that we don't want.'[164] In fact, GDF11 is currently being tested in clinical trials in aged humans. Whatever the outcome as an anti-ageing treatment, here is a clear example of a transcription factor involved in a complex bodily process being influenced by signals from the rest of the body. But this raises another important question—how exactly do proteins like CAP and CREB exert their effects upon genes?

Remarkably, pioneers like Jacob and Monod worked out the basic roles of such proteins as the lac repressor and CAP without any knowledge of how these proteins physically interact with genes.[140] But as the 1960s drew to a close and flower power and the Beatles' psychedelic phase were succeeded by protests against the war in Vietnam and the violent assassinations of Malcolm X, Martin Luther King, and Bobby Kennedy, a new generation of scientists had decided the way forward lay in finding ways to directly isolate such transcription factors and characterize them. In fact, the struggles on the street and in the lab were not always so distant, as shown when Jonathan Beckwith, already mentioned as the co-discoverer of the CAP protein, in 1969 found a way to cut out the DNA composing the lac operon, and thus became the first person to physically isolate a gene.[165] For this breakthrough Beckwith was awarded the Eli Lilly Award in Microbiology, but in the spirit of the time he gave the $1,000 prize money to the radical Black Panther Party.[166] He also voiced publicly his concerns that such 'genetic engineering' might have a more sinister side, either in the creation of new, deadly pathogens, or as a tool of a repressive state. However, while there were many similar arguments in the first years of the molecular biology revolution, they did nothing to stop the growth of what was becoming a technological juggernaut. For new techniques were making it possible to do things Watson and Crick could only have dreamed about in 1953.

Two young scientists who decided to isolate and characterize a transcription factor for the first time were Walter Gilbert and Mark Ptashne, both junior members of the same department at Harvard University, where Watson had taken up a professorship after the discovery of the double helix. Watson once said one recipe for successful science was to 'take young researchers, put them together in virtual seclusion, give them an unprecedented degree of freedom and turn up the pressure by fostering competitiveness'.[167] This was precisely the situation that developed in his own department when Gilbert and Ptashne decided to independently isolate a transcription factor. While nominally working towards quite separate goals—Gilbert focused on the lac repressor, while Ptashne tackled the repressor in phage—the situation rapidly turned into a race to see who could get to his goal first. At the time, Ptashne was heavily involved in the movement against the Vietnam War and even took part in a lecture tour of North Vietnam, in which he talked both about his scientific work and political activities even while the bombs were falling on the country. Horace Judson, who interviewed him at this time, noted his 'aviator-style spectacles, T-shirt, sawed-off blue-denim shorts, and sandals—more exposed skin than appeared prudent in a laboratory'.[168] But it was Ptashne's comment that 'people who claimed to be trying to isolate the repressor…weren't really willing to take the kind of risks that were necessary… *psychic* risks', that best gives a flavour of the highly-charged atmosphere in many molecular biology labs at this time. Meanwhile, for Gilbert, the repressor had become 'a holy grail…like isolating the neutrino…those of us who were involved in the isolation, of course, believed in its existence in a way other people did not'.[169]

Perhaps it was not so surprising that sheer force of will and an almost mystical faith in the likelihood of success were seen as necessary attributes for isolating a transcription factor, given that Monod's own calculations had indicated there were probably only seven or eight molecules of the lac repressor in a bacterial cell, or less than two thousandths of 1 per cent of the cell's protein.[169] Over the next few years, both Ptashne and Gilbert tried different biochemical strategies, use of radioactive labelling, and comparison of the properties of normal versus mutant bacteria, that led frustratingly close to their goals but then saw it vanish in a puff of smoke. And in the testosterone-fuelled environment of the Harvard department, both individuals knew that one's success would be viewed as the other's failure. Thankfully, both Ptashne and Gilbert finally achieved their goals of isolating their respective repressor proteins in publications that appeared almost simultaneously.[170] Success had come to both researchers without a resulting nervous breakdown, although the experience led Watson, qualifying his earlier remark about competiveness, to reflect that 'it is better that one's competitors be in a different city, if not country. Having them in the same building is a small model of hell.'[171] Most importantly, isolation of these transcription factors opened the way to that of other gene regulatory proteins, and made it possible to properly investigate the specific ways in which these proteins influence gene activity. In particular, such studies showed that transcription factors have a specific affinity for particular DNA sequences at the start of the gene they control, and illuminated the way they influence gene expression.

As we've seen, a key aspect of understanding how DNA functions as the hereditary molecule was the realization that its four different bases act as letters in a linear code. According to this view, the different sizes and shapes of the four bases are irrelevant. But when it comes to transcription factors, these differences become very important, since, as complex 3D structures themselves, proteins can only interact with other molecules through such properties of size and shape. To do so, transcription factors bind in the grooves of the DNA double helix where they can recognize a specific sequence by the differences in the shapes of the bases. Indeed, recent advances in X-ray diffraction have made it possible to identify the precise molecular interactions between transcription factors and their DNA target at a detailed level of structure.[172] But how does the binding of such factors determine whether a gene is turned on or off?

We've seen how mRNA is produced by RNA polymerase. However, on its own the polymerase is inactive and needs contact with proteins like CAP or CREB to activate it. By binding to DNA sequences at the start of a gene, such proteins are brought into close proximity to RNA polymerase, and this contact is sufficient to activate the polymerase. That transcription factors are themselves regulated by intracellular signals explains how information from the environment can influence gene action. This is one reason why, despite different cell types containing the same genomes, the proteins produced in such cells are very different. So heart cells typically contain proteins that regulate their contraction, while liver cells contain those involved in the metabolism of foodstuffs. The proteins produced in such different cell types differ partly because the cells contain different transcription factors, but also because of different incoming signals relayed by second messengers.

We can see now why such an enormous change took place when a differentiated cell nucleus was transplanted in an egg with its own nucleus removed, as happened during cloning of Gurdon's frogs, or Dolly the sheep. While an udder cell only contains the transcription factors needed to switch on genes involved in breast cell functions, like those involved in producing milk, in the egg a transplanted nucleus is exposed to quite different factors, those geared towards allowing the combined sperm and egg genomes to produce all the proteins required to make a whole new organism.[173] This is not all that needs to change in the transplanted nucleus, however, for another important difference between bacteria and multicellular organisms is that our DNA also comes wrapped in proteins that regulate accessibility of genes to transcription factors.

As we've seen in Chapter 2, in the late nineteenth century Friedrich Miescher and Walther Flemming showed that DNA in chromosomes is associated with proteins. Flemming gave the name chromatin to this combination of chromosomal DNA and proteins.[174] In 1884 Albrecht Kossel, who had studied alongside Miescher, showed that the main protein component of chromatin was a single type of protein, which he called histone.[174] Kossel also showed that histone was a positively charged, basic protein, explaining its affinity for negatively charged, acidic DNA. We've seen that one reason it took so long for DNA to be accepted as the molecule of inheritance was its supposedly boring structure, which was revealed when Kossel identified the four nucleotide bases in DNA and RNA.[174] Instead, it seemed more likely that histone would somehow specify the genetic information; however, with the recognition of the importance of DNA in the 1950s, interest in histones slipped away. Ironically, it was the protein in chromatin, not the DNA, that was now seen as boring.

Such a view did not really change until the 1970s, when scientists began to scrutinize chromatin structure in much greater detail. In 1974, Donald and Ada Olins of the University of Tennessee studied chromatin under the electron microscope (the only microscope with sufficient resolution to visualize DNA directly because electrons have a much shorter wavelength than visible light) and made an exciting discovery.[175] We've seen how one view of genes is that they are analogous to beads on a string. But now the Olins saw that chromatin really did look like a string of beads; albeit with each bead covering a far smaller portion of DNA than a gene. Further studies by Roger Kornberg showed that the 'beads' were in fact an octamer of four pairs of different histone subtypes—H2A, H2B, H3, and H4—around which a segment of DNA coils to form what Kornberg called a 'nucleosome' (see Figure 11). Kornberg was the eldest son of Arthur Kornberg—the discoverer of DNA polymerase—and he made his own important discovery while working as a young postdoctoral scientist with Aaron Klug and his colleagues at the Laboratory of Molecular Biology in Cambridge. Klug had carried out his PhD with Rosalind Franklin at Birkbeck College, London, using X-ray diffraction to study the structure of viruses, a topic which Franklin had turned to following her work on DNA.[176] The expertise Klug gained at this time allowed him and his team to analyse the structure of the nucleosome in fine detail. These studies revealed that DNA is coiled just over two times around the histone octamer, while another histone, H1, remains outside this core and is only found in

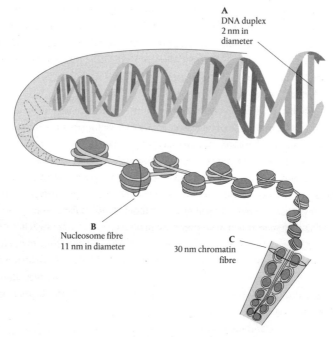

Figure 11. Basic nucleosome structure of chromatin

particularly tightly packed DNA.[177] Klug's estimation of his work was that it was 'not necessarily glamorous, nor does it often produce immediate results, but it seeks to increase our basic understanding of living processes'.[178] Certainly, its significance was viewed as sufficiently important by the Nobel Prize committee, who awarded the prize to Klug in 1982. Kornberg would receive his own Nobel Prize in 2006 for working out the fine detail of the process of transcription in eukaryotes—species whose cells have a nucleus, which includes complex multi-cellular organisms like ourselves, but also the unicellular yeast.

Initially, histones were thought only important for packaging DNA into a manageable form in the nucleus. However, in the 1990s, Michael Grunstein at the University of California and David Allis at the Rockefeller University, New York, showed that addition of an acetyl group (chemical structure $-COCH_3$) to histones by the cellular machinery radically alters their interaction with DNA.[179]

Addition of the acetyl group loosens the histone–DNA association, while removal tightens it. Chemically, this makes sense, since the acetyl group removes a positive charge from the histone, making it less attractive to negatively charged DNA. And the enzymes—histone acetylases and deacetylases—that mediate these changes, activate or repress gene expression respectively, by making DNA regulatory elements more or less accessible to transcription factors.[179]

Importantly, this discovery explained why the process of differentiation whereby 'totipotent' cells change into specialized cell types in a multicellular organism, is generally a one-way process. Only in exceptional circumstances, as during cloning, can such changes be reversed.[173] So the egg reprogrammes gene expression not only by providing a new set of transcription factors but also by making DNA control elements accessible to such factors. Despite these differences between gene expression in bacteria and multicellular organisms, one could still argue that they share much in common. So in both cases genes are switched on or off in response to incoming signals mediated by transcription factors that bind at the start of the genes they control. It was this that led Monod to claim 'what is true for *E. coli* is also true for the elephant'.[180] But as studies of gene regulation in multicellular organisms began to gather pace in the 1970s, other surprises were waiting on the horizon.

4

THE SPACIOUS GENOME

'It is a remarkable fact that the greater part (95 percent in the case of humans) of the genome might as well not be there, for all the difference it makes.'

Richard Dawkins

'Trying to read our DNA is like trying to understand software code—only with 90% of the code riddled with errors. It's very difficult in that case to understand and predict what that software code is going to do.' *Elon Musk*

The term 'survival of the fittest' generally conjures up images of lean, mean fighting machines, with lions, tigers, and great white sharks springing to mind. But really the only true measure of evolutionary success is an organism's ability to pass its genes on to the next generation. In this respect the humble bacterium is a clear winner. For bacteria have mastered the art of thriving in practically any environment on Earth, whether that be boiling hot springs in Yellowstone National Park, a dark and freezing lake deep under the Antarctic ice, or the confines of a human intestine.[181] Whether measured by cell number or sheer biomass, bacteria outperform every other life form on our planet. And a recent study that identified bacteria living happily a kilometre and a half below the Earth's crust shows the reach of these tiny organisms is even greater than suspected.[182] Such bacteria rely not on sunlight but on chemicals released by the rocks themselves. Bacteria have thrived over the three and a half billion years they have existed upon the planet, partly because of their small size and relative simplicity, which allows rapid reproduction and evolution, but also due to an ability to focus on one or a few sources of energy and exploit them as ruthlessly as possible.[181]

Bacterial genomes are admirably suited to this goal. The study of the lac operon showed that its genes are not only switched on simultaneously but transcribed into a single mRNA molecule.[183] Subsequent studies showed that other bacterial genes with a similar function are linked in this way. An initial assumption was that a similar situation would exist in multicellular organisms, including humans. But as such genes were further investigated it became clear that they were transcribed into RNA as single, not multiple, entities. In addition, these genes were not adjacent; indeed, they were often located on completely different chromosomes. In other respects, though, gene regulation in bacteria and multicellular organisms seemed initially very similar. Jacob and Monod gave the name 'promoter' to the region where RNA polymerase but also transcription factors like CAP bind, and as researchers began studying gene regulation in multicellular organisms they found a similar arrangement of transcription factor binding sites at the start of genes. However, it soon became clear that such short-range influences were only part of the story. In particular, gene regulatory regions named 'enhancers', due to their potency in boosting gene expression, seemed unlike anything identified in bacteria.

Enhancers were first identified independently by Pierre Chambon at Strasbourg University and George Khoury at the US National Cancer Institute. Both scientists were studying animal viruses in the 1970s, because a popular idea at the time was that viruses were a primary cause of cancer. Remarkably, the link between viruses and cancer was identified over a hundred years ago, in 1911, by Peyton Rous, a pathologist at the Rockefeller University, New York.[184] Investigating the cause of tumours in farm poultry, Rous discovered that a cell-free tumour extract could cause cancer in chickens into which it was injected.[184] However, few people believed his claim that this showed cancer could be caused by an organism even smaller than a cell, namely a virus, and it was only after others confirmed his findings in the 1950s that Rous was finally awarded a Nobel Prize in 1966.

In fact, viruses only cause cancer in a minority of cases, such as certain types of human cervical cancer linked to infection by papillomaviruses. The discovery of this link by Harald zur Hausen of Heidelberg University has led to vaccination of teenage girls in some countries as a preventative measure, and to zur Hausen being awarded a Nobel Prize in 2008.[185] However, the focus on animal viruses in the 1970s also resulted in major insights into gene regulation in multicellular organisms, since such viruses reproduce by hijacking their host's own gene expression machinery. As such, because of their relative simplicity, studies of viruses offered an indirect way to investigate gene expression in multicellular organisms.

Enhancers were discovered by Chambon and Khoury while independently investigating which regions of viral genomes are most important for their regulation. Surprisingly, they found that some viral DNA sequences retained their potency even when thousands of base pairs away from the genes they regulated, and seemed equally capable of affecting a gene's activity whether located before or after it.[186] At first, enhancers were thought to be just a quirk of viruses, but then other studies showed that regulatory sequences acting many kilobases distant from the genes they controlled were also key features of human genes.[187] All this was most perplexing, for it completely contradicted the idea of genes, and their regulatory regions, as being compact units as established in bacteria.

If Jacob and Monod's findings had suggested that genes were like workshops, each producing its own specific product, on an incredibly long road—the chromosome—it now seemed that, in multicellular organisms, the on/off switch for each workshop could be miles down the road, with no obvious physical connection to the object it controlled. So how was such activity at a distance possible given that transcription factors were thought to directly interact with the RNA polymerase? One initial idea was that proteins binding to enhancers triggered a change in the DNA helix that was transmitted to the start of the gene. But how this might work remained unclear; instead, subsequent studies suggested that transcription factors bound to enhancers activate the RNA polymerase directly, by looping around the intervening DNA separating them (see Figure 12).[187] While enhancers activate genes, other DNA regions acting at a distance have the opposite effect, earning them the name of silencers.

One important feature of enhancers is their flexibility. Recently, Robert Tjian of the University of California has suggested that this flexibility may have been key to the development of complex multicellular life forms. An amazing aspect of multicellular life, whether a human or a fruitfly, is how a single cell—the fertilized egg—subsequently becomes an exquisitely structured organism with multiple types of cells, tissues, and organs, all working together in harmony. However, this creates a huge challenge, for not only must the individual cells of a developing multicellular organism respond to changes in their immediate environment, as a bacterium does in utilizing available nutrients, but regulatory mechanisms during embryogenesis must also be structured so that gene expression is tightly coordinated across the body.

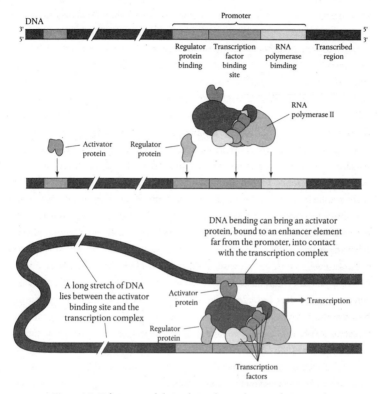

Figure 12. Enhancers and their relationship to the gene they control

One basic way in which an organism is structured is along its body axes.[188] Like other vertebrates we humans have a top and bottom, a back and front, plus a left and right side. Vertically, we have a head containing a brain at the top of our bodies, a torso with two arms at the upper end and two legs at the lower end, while in the other direction we have a back with a spine and shoulder blades and a front with chest and midriff. Finally, our internal organs are positioned according to a left–right asymmetry.[189] Most people's hearts are on their left side, although exceptions to this rule exist, and not just fictional characters like James Bond's adversary Dr No, who survived an assassination attempt because his heart was on his right.[190] So singer Donny Osmond, whose picture adorned the bedroom walls of teenage girls across the world in the 1970s, also has his internal organs reversed in this way. He only found this out following a bad case of appendicitis that was initially misdiagnosed because his appendix was on the left instead of its normal right-hand position.[191] At a finer level of structuring, humans typically have five digits on their hands and feet, although people can be born with an abnormal number: the world record holder for the most digits being Akshat Saxena, an Indian boy born in 2010 with seven digits on each hand and ten on each foot.[192]

There has been scientific interest in the genes underpinning body axes for many years, partly reflecting a desire to understand the basic mechanisms of embryo development, but also because some abnormalities in human body 'patterning' pose serious threats to well-being and survival. A major step forward in our understanding of the genetics of body patterning came with the discovery of mutants affecting the body plan in fruitflies. The first such mutant was identified by Calvin Bridges in Thomas Morgan's laboratory in 1915, but it was Ed Lewis who first studied such mutants in detail. Lewis did his PhD with Alfred Sturtevant, the creator of the first genetic map, at the California Institute of Technology, where Lewis himself later established his own laboratory. It was here he discovered various mutants whose body symmetry was out of sync, such as flies with legs on their heads, or four instead of two wings.[193] Subsequent studies by Christiane Nusslein-Volhard at Tübingen University identified the genes linked to such bizarre mutants, a discovery that led to the award of a Nobel Prize to her and Lewis in 1995. These 'homeotic' genes, from the Greek for assimilation, code for transcription factors that switch on other genes, establishing a hierarchy of gene regulation, with genes involved in gross body patterning controlling those regulating a finer level of detail.[193] Remarkably, much the same genes that define different structures along the length of a fly are responsible for patterning our own human bodies.

One such set involved in body patterning are known as the pair-rule genes. These genes are expressed as seven stripes along the length of the fly embryo, which define where the different structures of the adult organism will later appear.[194] This complex pattern of expression is regulated by multiple enhancers, each specific for a different stripe, since if one is disabled by genetic engineering, this causes the loss of the stripe of expression it controls.[194] This provides a potential mechanism for the evolution of complex multicellular life forms, for by increasing the enhancers controlling a gene, each enhancer controlling the gene's expression in a different part of the embryo, evolution could first generate a simple segmented structure, as in worms, and, subsequently, more complex structures, like limbs, wings, and so on.[195] This required, however, a dramatic loosening of the connection between a gene and its regulatory elements, in contrast to the tight link in bacteria.

But it isn't just the regulation of gene activity that is radically different in multicellular organisms, but the very structure of the genome itself. The first indication of this came from comparisons of genomes from different multicellular species. Such analysis identified huge discrepancies in the sizes of certain genomes compared to the apparent complexity of the species they came from.[196] The fact that some organisms had far more DNA than humans posed a potential threat to the view of ourselves as a superior species, if a greater amount of DNA was assumed to represent a more complex organism. Another possibility, though, was that the genomes of multicellular organisms contained an excess of non-functional 'junk' DNA, which varied between species. This possibility was bolstered by the first studies to examine the actual sequences of DNA in our genomes. In the 1960s, in the absence of any method for directly 'reading' the sequence of bases in DNA, Roy Britten and David Kohne at the Carnegie Institution in Washington, developed an ingenious way to do this indirectly.

If DNA is heated, eventually it acquires sufficient energy that the two strands of the double helix come apart. If the DNA is then cooled, because of the attraction of bases on the two strands for each other, eventually the double helix reassociates. Because it takes time for any piece of DNA to find its complementary sequence, Britten and Kohne assumed that the mouse genome would take much longer to reassociate than that of bacteria, given the latter's 100 times smaller size. Instead, they found that the reassociation occurred in waves: a quarter doing so rapidly, a further third more slowly, and only the remainder combining at the slow speed one would expect. The two researchers concluded that the more rapidly reassociating portions of the mouse genome must be highly repetitive, since such sequences are so similar they need not find their exact partners. To relate such genomic regions to a role in coding for proteins, Britten and Kohne next incorporated radioactivity into mouse mRNA and included this in the experiment. This showed that the mRNA only matched the slowest reassociating portion of the genome. It therefore seemed that over half the genome was made up of repetitive sequences, these being also 'non-coding' regions. This discovery would play an important part in the rise of the belief that protein-coding genes are islands in a sea of 'junk' DNA, although Britten and Kohne themselves found this idea 'repugnant', preferring to believe that the function of repetitive DNA simply hadn't been identified.[197]

Further surprises were in store once scientists began to look more closely within genes themselves. As we've seen, while DNA acts as the ultimate repository of hereditary information, mRNA is the actual template for the assembly of an amino acid chain. As such, it seemed fair to assume that mRNA would be a direct replica of the DNA in order to fulfil this role. In bacteria this was found to be the case, with a one-to-one correspondence between the DNA and RNA molecules. But when scientists began similar comparisons in eukaryotes in the late 1970s, what they found was totally unexpected. The key discovery was made in 1977 by Phillip Sharp at the Massachusetts Institute of Technology and Richard Roberts at Cold Spring Harbor Laboratories.[198] While studying the reproductive cycle of adeno-viruses—which cause human illnesses ranging from the common cold to bronchitis and pneumonia—both were intrigued to find that adenoviral RNAs in the nucleus of the infected cell were much bigger than those in the cyto-plasm. Although at this time it was impossible to sequence DNA and RNA routinely, a comparison could be made by allowing them to bind to each other by the same 'base-pairing' that occurs in a DNA double helix. Studying such DNA–RNA hybrids under the electron microscope, a curious sight was observed: at regular intervals along the hybrid, large loops were seen. Further analysis identified these as DNA, indicating that only a small proportion of the DNA sequence in the gene was present in the mRNA. Curiously though, RNA from the nucleus was much longer and it perfectly matched the DNA of the gene. This suggested that, initially, an RNA spanning the length of the gene was produced but then it was substantially trimmed to size. This phenomenon was named splicing, by analogy with the way footage for a film is shot, with sections being cut out to create the final product (see Figure 13).[198] Rather confusingly, the discarded regions were named introns, for intragenic regions, and those that remained in the final mRNA were called exons, since these were expressed as protein. Any idea that splicing was some quirk of viral gene expression was soon quashed by further studies that showed it was also a feature of various human genes, such as the immunoglobulin genes and the globin genes that code for antibodies and haemoglobin respectively.[199]

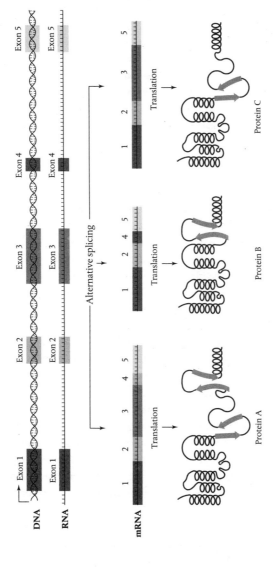

Figure 13. RNA splicing and alternative splicing

In fact, studies have shown that, in multicellular organisms, genes without introns are very much the exception, as well as highlighting the huge discrepancy in size between introns and exons.[199] So while the normal length of exons is less than 200 bases, introns can be anywhere from 2,000 to 11,000 bases long.[200] This disproportion in size is such that if the genome were a book, with the genes as different chapters, such chapters would contain chunks of only a paragraph or so of meaningful text, interspersed by pages of gibberish. So why has evolution allowed the creation of such long RNAs, if so much is subsequently thrown away? One reason may be the extra flexibility to the organism that can result. In different cell types of the body, or at specific stages of embryo development, different exons can be selected to be included in the final mRNA. This 'alternative splicing' means a single gene can code for many different proteins (as seen in Figure 13).[201] While there is still a flow of information from DNA to RNA to protein, this ensures that such information can go in different directions depending on the cellular environment. Alternative splicing can alter a protein's mode of action, regulation by cellular signals, or interaction with other proteins, to highlight just some ways in which a protein's function can be altered by this process. Its importance in humans is shown by the fact that over 90 per cent of our genes are alternatively spliced.[201]

Alternative splicing plays a particularly important role in the formation of antibodies by the immune system. It generates antibodies against a seemingly unlimited range of different foreign molecules, or antigens. Initially, there seemed two possible explanations for this, one being that when the body comes into contact with a foreign antigen it generates a specifically tailored antibody to match, an alternative being that it generates an almost infinite variety of different antibody molecules.[202] Using an analogy of buying a suit, these alternatives are like having one made-to-measure from a bespoke tailor, or buying a ready-made item from a high street clothing chain. Initially, the former model seemed most plausible, for how could the body generate such a huge variety of different forms of a protein? Yet this is indeed what happens, with alternative splicing being one mechanism whereby a single antibody gene can code for many different protein forms.[203]

Another reason why splicing might have been favoured in multicellular organisms are its potential benefits for evolutionary change. A crucial aspect of Darwin's theory of natural selection as a mechanism of evolution is its reliance on blind chance. It is this aspect of Darwinism that can seem most threatening to those who look for some kind of guiding influence to life. In Darwin's own lifetime his theory was derided as 'the law of the higgledy-piggledy' by astronomer and philosopher John Herschel. This was ironic, given that Herschel himself probably stimulated Darwin to start thinking about evolution when they met at the former's home in Cape Town during the voyage of the *Beagle*.[204] For Herschel was sympathetic to the idea that new species could come into existence, which he called the 'mystery of mysteries', and indeed Darwin directly referred to this phrase and to Herschel in the *Origin of Species*. However, what the latter had in mind was a 'directed' evolution administered by God. Herschel also influenced Darwin in his general approach to science, through his statement that scientific discoveries are made when the mind 'leaps forward... by forming at once a bold hypothesis'. Unfortunately, Darwin's great idea was rather too bold for Herschel in the fact it went beyond conventional accepted notions.[204]

Others have argued that the chance of complex organisms such as human beings evolving is as likely as a tornado blowing through a junkyard, reassembling the dismembered remains of a Boeing 747 back into a fully functioning Jumbo Jet. This analogy, first used by the astronomer Fred Hoyle,[205] is based on a deep misunderstanding of natural selection: the variants available may arise by chance, but the actual selection, and therefore development of adaptations, is not arbitrary at all but moulded by the environmental conditions. However, it also neglects another feature about evolution, namely its conservatism, never creating anything purely from scratch but always borrowing from what is already there. And, in this respect, the division of eukaryotic genes into intron and exon regions seems to have played a very important role. So studies of proteins have shown these are composed of 'domains', each with their own discrete 3D structure separate from the rest of the protein.[206] Moreover, these structural domains often have a discrete function within a protein. Intriguingly, the same domains may crop up in proteins of quite different overall functions. It seems that, during evolution, new proteins have formed by mixing and matching existing domains. That protein domains often map onto specific exons, and are therefore already separated by introns at the DNA level, has been a key factor in allowing this process to occur: what is known as 'exon shuffling' (see Figure 14).[206]

Not that splicing doesn't have its disadvantages. In particular, serious genetic disorders can result from errors in the process.[207] One such disorder is Tay–Sachs

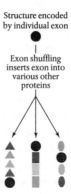

Structure encoded
by individual exon

Exon shuffling
inserts exon into
various other
proteins

Figure 14. Exon shuffling as a central mechanism of protein evolution

disease, named after clinicians Waren Tay and Bernard Sachs, who, in the late nineteenth century, first noticed its occurrence among children of Ashkenazi Jewish immigrants in the US.[208] Symptoms begin as early as 6 months of age, when a previously normal child's development begins to slow, followed by rapid weakening of the muscles, loss of vision and hearing, and eventually full-scale dementia. Tragically, those with the condition die by the age of 3 to 5 years old. We now know this devastating disease is caused by a defect in the enzyme HexA, which normally breaks down a fatty substance called GM2 ganglioside.[209] In HexA's absence this substance builds up in cells of the nerves and brain, causing them to stop working normally and eventually destroying them. Tay–Sachs is a Mendelian recessive disorder, and, as such, is passed on by two carriers who do not themselves suffer from the disease. In one common form of Tay–Sachs, a mutation in an intron–exon junction means that the intron fails to be excised and so remains in the final mRNA where it disrupts the protein code, leading to a dysfunctional enzyme. Unfortunately, despite our detailed knowledge of the genetic basis of Tay–Sachs, this devastating disease remains incurable. However, there have been important steps forward in its prevention. This is mainly due to one man, Rabbi Josef Ekstein, who, having lost four of his children to the disease, set up a premarital testing service for potential couples in the Ashkenazi Jewish community, in which arranged marriages still play an important role.[210] If both individuals test positive they are told that the marriage cannot go ahead. This service has drawn some criticism on ethical grounds but its success at disease prevention cannot be doubted, since the incidence of Tay–Sachs among Ashkenazi Jews is now less than in the general population.[208]

Such a link between splicing and disease was still some way in the future when splicing was first discovered. At that time, its main significance was providing more evidence of how different our own genomes are compared with those of bacteria, and how even genes themselves seem full of non-coding junk. But the full extent of non-coding DNA became most evident once the human genome itself was sequenced. The sequencing technique used was developed in 1977 by Fred Sanger, who as we saw in Chapter 2, two decades earlier, had obtained the first protein sequence. Sanger has referred to his career after successfully sequencing insulin as his 'lean years';[211] however, as his efforts resulted in him developing ways to sequence first RNA, then DNA, and the award of a second Nobel Prize, in this case leanness is a relative concept! Having tried numerous unsuccessful approaches in his bid to sequence DNA, he finally succeeded with one involving DNA polymerase, the enzyme that replicates DNA. Sanger realized that synthesis of a DNA strand from a template DNA could be used to 'read' the sequence of that template, if some way were found to interrupt DNA elongation in a manner specific to each base. He did this by generating modified nucleotides that were incapable of being linked to a subsequent nucleotide, so acting as a 'chain breaker'.[211] By having four tubes, each with a modified nucleotide corresponding to one of the four bases, a mixture of different lengths of DNA were obtained, each ending at either A, C, G, or T, depending upon the tube. By also radioactively tagging the DNA and separating it with a technique called gel electrophoresis, it was possible for the first time to 'read' the DNA sequence (see Figure 15). Sanger received a Nobel Prize for this discovery in 1980, making him one of the few people to have received the award twice.

Crucially, a refined version of Sanger's method, in which the four modified nucleotides each impart a different colour to the DNA and allow the process to be carried out in a single tube, made it possible to automate DNA sequencing. This was the approach used for the genome project.[212] The findings of this vast sequencing project were surprising in a number of different ways. The first surprise was the number of protein-coding genes in our genomes. A common estimate of the number of human genes prior to the genome project was 100,000. But the real figure is far lower, with a recent study quoting it as just over 22,000. This is more than a fruitfly, with 15,000 genes, or a chicken, with 17,000, but less

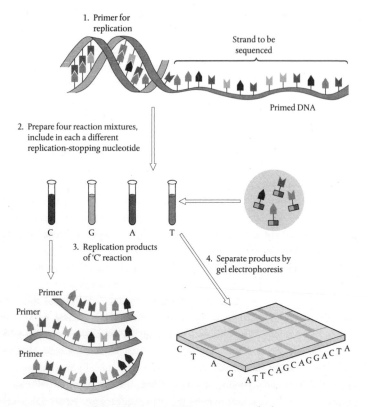

Figure 15. Sanger's DNA sequencing method

than a grape, with over 30,000 genes (see Figure 16). So much for the genetic superiority of our species, at least as assessed by gene number alone. But the other big surprise was how little of our genomes are devoted to protein-coding sequence. So when the DNA present in our introns and between genes is compared to that coding for proteins, a seemingly insignificant 2 per cent is devoted to the latter. This finding greatly strengthened a claim made decades earlier by Susumu Ohno of City of Hope Medical Center in California, when he said, in 1972, that most of the human genome was what he called 'junk' DNA.[213] In this

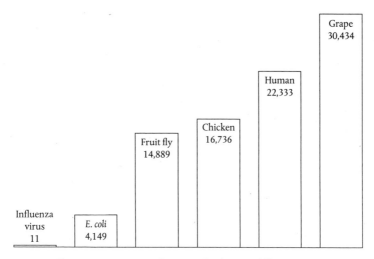

Figure 16. Comparison of gene number between different species

first use of the phrase, Ohno was referring to 'pseudogenes', namely genes that originally resulted from the duplication of a functional gene, but which have become disabled through mutation, so they no longer produce a functional protein. However, this term soon became a popular way of referring to all the non-coding DNA in the genome.

At first glance, the existence of junk DNA seems to pose another problem for Crick's central dogma. If information flows in a one-way direction from DNA to RNA to protein, then there would appear to be no function for such non-coding DNA. But if 'junk DNA' really is useless, then isn't it incredibly wasteful to carry it around in our genomes? After all, the reproduction of the genome that takes place during each cell division uses valuable cellular energy. And there is also the issue of packaging the approximately 3 billion base pairs of the human genome into the tiny cell nucleus. So surely natural selection would favour a situation where both genomic energy requirements and packaging needs are reduced fiftyfold? An influential explanation for how the majority of a genome could become junk rather than being eliminated by natural selection was put forward by Richard Dawkins of Oxford University in 1976 in his book *The Selfish Gene*. In this, he noted that 'if the "purpose" of DNA is to supervise the building of bodies, it is surprising to find a large quantity of DNA which does no such thing. Biologists are racking their brains trying to think what useful task this apparently surplus DNA is doing...The simplest way to explain the surplus DNA is to suppose that it is a parasite, or at best a harmless but useless passenger.'[214] Francis Crick and Leslie Orgel later put this idea on a more formal scientific footing in 1980, in an article in which they presented evidence that such junk represented parasitical DNA, with an ability to reproduce itself but whose accumulation did not have enough of a detrimental effect on an organism's physiology or behaviour to allow it to be eliminated by natural selection.[215]

The existence of junk DNA has been proposed as further evidence that humans evolved by natural selection rather than being created by some supernatural being. In 1802, the theologian William Paley used the existence of an exquisitely complex biological structure like the human eye as evidence for God, saying 'Is it possible to believe that the eye was formed without any regard to vision?... Design must have had a designer.' [216] Darwin saw the human eye as a particular challenge for his theory, since its many sophisticated features seem interdependent, posing problems for his stress upon the power of gradual step-by-step change to transform life. His answer was to point to organisms with eyes ranging from simple to complex, and to suggest that evolution of the human eye involved similar organs as intermediates.[217]

But what if design is not always so perfect? Harvard palaeontologist Stephen Jay Gould believed that nature's oddities such as the panda's thumb—actually an enlarged wristbone, and which Gould saw as a rather clumsy solution to a design problem—serve as better proof of evolution's existence than more 'ideal' adaptations.[218] In this sense, what more perfect demonstration is there that nature is 'an excellent tinkerer, not a divine artificer', than the fact that 98 per cent of our own genome is useless? Certainly, this is an argument Dawkins has employed as evidence against a religious interpretation of life's origins, saying that there is no 'convincing reason why an intelligent designer should have created a pseudo-gene—a gene that does absolutely nothing and gives every appearance of being a superannuated version of a gene that used to do something—unless he was deliberately setting out to fool us'.[219] Similarly, Kenneth Miller of Brown University has observed that the genome resembles 'a hodgepodge of borrowed, copied, mutated, and discarded sequences and commands that has been cobbled together by millions of years of trial and error against the relentless test of survival.

It works, and it works brilliantly; not because of intelligent design, but because of the great blind power of natural selection.'[220]

This is a powerful argument, and one that I have much sympathy with, guided as I am by the principle that both life and the universe can be explained by purely materialist principles. However, employing the uselessness of so much of the genome for such a purpose is also risky, for what if the so-called junk turns out to have an important function, but one that hasn't yet been identified? Whether such important functions exist within non-coding DNA has been one of the most hotly debated topics in genetics over the last few years. And one way in which this question first began to arise was through a reconsideration of the role of DNA's chemical cousin, RNA.

5

RNA OUT OF THE SHADOWS

'Because all of biology is connected, one can often make a breakthrough with
an organism that exaggerates a particular phenomenon, and later explore the
generality.'   *Thomas Cech*

'As is a frequent occurrence in science, a general hypothesis was constructed
from a few specific instances of a phenomenon.'   *Sidney Altman*

Trying to conjure up the past is never easy, especially when that past is 3.7 billion
years old. In his *Origin of Species*, Darwin was noticeably cagey about the precise way
in which he believed life first arose on Earth, stating only that 'probably all the
organic beings which have ever lived on this earth have descended from some
primordial form, into which life was first breathed'.[221] However, writing in private
to his friend the botanist Joseph Hooker, he was prepared to be more speculative.
'But if (and oh! what a big if!),' he wrote, 'we could conceive in some warm little
pond, with all sorts of ammonia and phosphoric salts, lights, heat, electricity etc.
present that a protein compound was chemically formed ready to undergo still
more complex changes.'[222] This showed a recognition that the conditions that first
gave rise to life might be very different to those in our current world. Meanwhile,
the focus on proteins as central to life's origins reflected the idea, even at this time,
that these molecules were key mediators of bodily processes. However, a more
precise suggestion for the type of chemical environment likely to have existed on
primeval Earth came in the 1920s from Russian biochemist Alexander Oparin, and
J. B. S. Haldane.[223] Both independently proposed that our planet's early atmosphere was likely to have consisted of methane, ammonia, carbon dioxide, and
water; with energy supplied by volcanic eruptions or lightning, this ought to have
been sufficient to generate amino acids and other building blocks of life.

In 1953 Stanley Miller and Harold Urey at Chicago University first explored, experimentally, the possibility that such a mixture of simple chemicals could give rise to more complex molecular structures.[222] By exposing a mixture of water, methane, ammonia, and hydrogen in a sealed flask to an electric spark—to mimic lightning—and heat—to stimulate continual evaporation and condensation— they found that, within two weeks, amino acids had formed.[224] With the dis- covery that DNA acted as a linear code for the production of proteins, emphasis shifted to showing that its building blocks too could have been generated in a primeval soup, and, indeed, subsequent experiments with a slightly different starting mixture showed this was the case.[222] However, there was now a major conundrum to be solved in explaining how a DNA template coding for protein production could have come into existence. On the one hand, such a template is the repository of the information in the cell, but it is also relatively inert. Indeed, it is this inertness that makes DNA ideally suited to its role as genetic material, and the reason why it has recently proven possible to extract information from the DNA of Egyptian mummies, Neanderthal fossils, and woolly mammoth tissues revealed by melting glaciers.[225]

Proteins, on the other hand, are highly active, and easily degraded. However, as we saw in Chapter 3, without them the information in DNA would mean little, for its code can only be 'read' by proteins like RNA polymerase, that transcribe the DNA code into its RNA intermediary, and other regulatory proteins, that activate the polymerase. Yet this presents a 'chicken and egg' situation, for if DNA can only be replicated with the help of catalytic proteins, but such proteins can only be propagated through a DNA code, then how could either arise on its own?[226] An important clue to solving this conundrum emerged in 1981, from studies of the process whereby mRNA is generated by splicing, and is subsequently used as a template to produce a protein, so-called 'translation'. Ribosomes—the subcellular machines that take a particular mRNA and use it to produce a protein correspond- ing to the genetic code contained in the mRNA—are composed of ribosomal proteins but also of ribosomal RNA, or rRNA for short.[227] In fact, the ribosome is highly complex, containing almost a hundred different proteins and a variety of different RNAs. Studies of the spliceosome showed that it too was a complex structure composed of proteins and RNAs.[227]

When scientists first began studying the mechanism of action of ribosomes and spliceosomes it was assumed that the proteins would be responsible for catalysis, with the RNAs performing an essentially structural role. This was an understandable assumption given that all the activities in life were believed to be carried out by special kinds of proteins called enzymes. These were first identified in 1879 by Eduard Buchner, who showed that yeast extracts lacking any living cells could still carry out the process of fermentation.[228] Then, almost fifty years later, in 1926, James Sumner crystallized an enzyme, urease, and showed it was a protein. We now know that enzymes are catalysts that allow the body's chemical reactions to take place in a fraction of the time they would require if left uncatalysed. Enzymes have a diversity of roles, digesting foods in the gut but transporting the digested products into the cell, transforming these into energy, and regulation of the genes coding for these processes.

However, when Thomas Cech of Colorado University began studying the molecular mechanisms underlying splicing in the early 1980s, he made a surprising discovery. Seeking to isolate and characterize the enzyme responsible for removal of the introns in rRNA in a single-celled eukaryotic organism called Tetrahymena, Cech found that the catalytic activity of the spliceosome was associated not with a protein, but a spliceosomal RNA.[229] Sidney Altman, at Yale University, was working on a different problem at this time—the generation of tRNAs in bacteria. His studies showed that such mature tRNAs are produced by a processing step, and, assuming this would be catalysed by an enzyme, he set out to isolate and characterize such an enzyme. But, like Cech, he found that the catalytic activity was due to an RNA.[230] Cech and Altman named such catalytic RNAs ribozymes, to stress the similarity with enzymes. Their suggestion was viewed with incredulity at first (you may be noting a pattern about great scientific discoveries by now), with some claiming that catalysis by RNAs was just a peculiar quirk of these systems. However, studies of the structure and mechanism of action of ribosomes—the protein production 'factories'—established the centrality of ribozyme action in the cell by showing that here too it was the rRNAs, not the ribosomal proteins, that constituted the ribosome's catalytic core.[231] The importance of Cech and Altman's discovery was acknowledged by the award of a Nobel Prize to them both in 1989. Altman later drew attention to two aspects of science, namely that 'hard work in stable surroundings could yield rewards, even if only in infinitesimally small increments', but also the emotional highs that could result from such work.[232] So, he recalled having 'resolved a problem that I had been working on for a year or more ... The feeling of great satisfaction at having solved my problem as well as having illuminated others kept me floating on air for weeks.'[233]

That RNA can function as a catalyst was initially puzzling, given that enzymes were thought to be the only molecules capable of forming complex 3D structures that provide specific catalytic pockets for the molecules they act upon. However, subsequent studies showed that RNA can form complex structures too.[234] We saw in Chapter 2 how DNA is a very uniform, some might say boring, molecule compared to proteins, with a chromosome's double helix structure the same from its start to its end. However, despite only having four types of bases like DNA, compared to the 20 amino acids in proteins, RNA can form complex 3D structures through the same base pairing that holds the DNA double helix together, but in a much more diverse manner than the latter. The discovery of ribozymes forced a reconsideration of RNA's role in the cell. Whether as a messenger or a component of the protein synthesis machinery, RNA had been relegated to a largely passive role in cellular function. But, with the discovery of its catalytic properties, speculation began as to what this meant for the molecule's function.[234] Was it possible, for instance, that RNA's flexibility of action reflected a past in which the molecule had played a much more central role in the replication of life than it now did? In fact, this possibility, named the 'RNA world hypothesis', was suggested as early as 1962 by Alexander Rich at the Massachusetts Institute of Technology, but his proposal had languished in the absence of evidence to support it.[226]

Now, with the discovery of ribozymes, the idea that on primeval Earth RNA had been both the molecule of inheritance and also the active motor of the cell, gained a new plausibility. According to this view, over time the more stable DNA usurped the role of RNA as the repository of life, while proteins increasingly took over the role of catalysis (see Figure 17).[226] Further support for the idea that RNA originally acted as the molecule of inheritance has emerged with the demonstration, in 2011, by Philipp Holliger at the Laboratory of Molecular Biology in Cambridge, that it is possible to artificially create a 'self-replicating' RNA molecule. This molecule not only replicates itself but can also generate another type of ribozyme, suggesting that once the first self-replicating RNA appeared, it might have generated a range of accessory molecular partners, kick-starting the evolution of more complex life forms. Another important clue as to RNA's past role comes from the fact that the second messengers cAMP and cGMP, and also the 'energy currency' of the cell ATP—whose chemical breakdown powers most catalytic

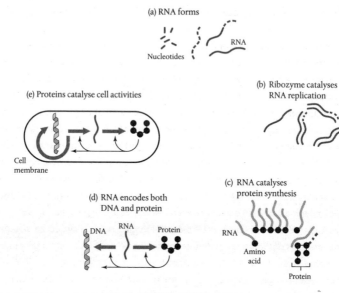

Figure 17. RNA world hypothesis

reactions—are all ribonucleotides or derived from these molecules, in line with a generalized role for RNA and its subsidiaries in the formation of early life.[226]

One important unresolved issue is how the origin of the first replication molecule relates to that of the first cell. We saw in Chapter 3 how the one-way flow of information from DNA to protein in Crick's central dogma ignores the fact that changes in the cellular environment, working through proteins that regulate gene expression, represent an important flow of information in the reverse direction. Similarly, discussion of life's origins has tended to assume the replicator arose first and then somehow acquired a cellular membrane. However, Jack Szostak at the Massachusetts General Hospital, who is 'interested in the related challenges of understanding the origin of life on the early earth and constructing synthetic cellular life in the laboratory', believes it more likely that a primitive cell formed first and then acquired replicator molecules, since it's hard to imagine how a replicating system could survive without a membrane to keep it from dispersing.[235] In line with this, recent studies suggest that primitive membranes could have been sufficiently permeable to allow important molecules within their orbit, but protective enough to retain them so as to allow the first life to develop.[236]

Of course, in the absence of a machine to go back 3.7 billion years in time it's impossible to really be sure what happened in our evolutionary past. And perhaps the only way we will ever know is if life is discovered elsewhere in our solar system, such as on Jupiter's moon Europa where chemicals such as water, ammonia, and methane—similar to those that gave rise to life on our own planet—are known to exist.[226] Returning to our own planet and its life forms, despite the realization that RNA catalysed some key processes in the cell, and had probably been a much more central player in life as a whole, at first this role was very much seen as past glory, with RNA representing, in Jim Watson's words, an 'evolutionary heirloom'.[237] But more surprises were on the way, and they initially came through an experiment that went wrong.

A popular idea about scientific discovery is that it is a highly logical process, in which scientists put forward hypotheses and then, by demonstrating their validity by experiment, gradually move towards a truer picture of the natural world. Although such patient deduction of the sort beloved by Sherlock Holmes is an important aspect of science, overly focusing on this underestimates the importance of luck in scientific discovery. In particular, experiments that don't give the desired result, but nevertheless reveal a view of the natural world not glimpsed previously, play a tremendously important role in science. One experiment that went wrong but led to a major insight was led by Richard Jorgensen, a plant scientist at the University of Arizona interested in the genetics of colour. He and his team tried to make a more intensely purple petunia by adding an additional pigment-producing gene to plants that were normally purple. However, instead of enhancing the colour, this genetic modification had completely the opposite effect, with the resulting flowers turning totally white, or becoming irregularly coloured.[238] As the normal and foreign genes seemed to cancel out each other's properties, Jorgensen and his colleagues called this phenomenon 'co-suppression'. Initially, the phenomenon was thought to be peculiar to petunias, but then other scientists started noticing similar results in other plant species. Subsequently, researchers studying the fungus Neurospora found that introducing additional copies of genes normally present in this organism silenced the effect of the normal genes.[239] But it still wasn't clear what caused the weird silencing effect, nor whether the phenomenon was confined to plants and fungi.

How silencing worked eventually emerged from studies focusing on the nematode worm *Caenorhabditis elegans*. This species was pioneered by Sydney Brenner, following his major contribution to cracking the genetic code, as a model organism for the study of embryo development.[240] By randomly mutating different genes with irradiation or mutagenic chemicals and then seeing whether such mutants had defects in embryogenesis, Brenner, and a growing army of 'worm specialists', began to identify many important genes involved in this process. One mechanism of development studied in detail was apoptosis, or 'programmed cell death'.[241] Although cell growth and division is a key aspect of embryogenesis, controlled cell death is also very important. Just as Michelangelo carved the statue of David from a single block of marble by hewing 'away the rough walls that imprison the lovely apparition to reveal it to the other eyes as mine see it',[242] so the worm embryo gains its detailed shape and form through trimming via cell death. Apoptosis also plays important roles during human development and in adult humans, for instance, in the destruction of cells that have become a health risk because their DNA is damaged.[240] Indeed, one way tumours develop is by ignoring the normal signals that trigger cell death. Sydney Brenner, who had never received a Nobel Prize for his work on the genetic code, finally did so in 2002, for discovering the mechanisms underlying apoptosis, which was made possible by the identification of worm mutants in which the process was defective.[240]

However, creating mutant nematodes is a laborious process, and some scientists wanted to find more direct ways of interfering with gene function. One such method was 'anti-sense' technology. This involved injecting a single strand of RNA complementary in its base sequence to a portion of a particular mRNA, the idea being that base pairing between the two would interfere with the mRNA's translation into protein. Andrew Fire and Craig Mello, at the University of Massachusetts Medical School, tried this technology in nematodes. They found that the anti-sense RNA did cause modest silencing of the gene, but, curiously, in another example of an experiment that went wrong, so did a control sense RNA, despite the fact it should not recognize the target mRNA by base pairing.[243,244] Even more surprising, when both anti-sense and sense RNA were added, this caused silencing that was a hundred times greater than with either component on its own. Because of this greatly enhanced effect, Fire and Mello realized the silencing agent must be double-stranded RNA, and the potency of the response suggested that the process was a catalytic one, since tiny amounts of such RNA caused silencing.[243] Mello has discussed how developing a new technology can be 'exceedingly frustrating because you may never know how close you were to success, and failures quite often teach you nothing. Partly because of this, those working on technology development often tend to band together and share ideas more than would otherwise be common among scientists. This was certainly the case for Andrew Fire and me.'[244]

The search for the catalytic agent led to the discovery of a multi-subunit protein complex, called the RNA-induced silencing complex, or RISC for short (see Figure 18).[245] This complex has two components, the first being a protein named DICER, because it chops double-stranded RNA into much smaller fragments called short interfering RNAs, or siRNAs for short. The second component is the Argonaute proteins, which attach themselves to the siRNAs and transport them to their matching sequence target in the mRNA, which is then inactivated. It now became clear that the previous observations of silencing had accidentally

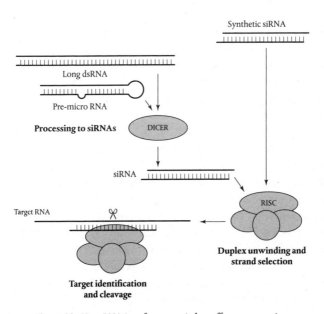

Figure 18. How RNA interference switches off gene expression

produced double-stranded RNA corresponding to a particular gene, which then activated RISC. According to Fire, 'we came into a field where a lot was already known. It was a complex jigsaw puzzle, and we were able to contribute one piece. Fortunately for us it was a very nice piece.'[246] He and Mello called the process 'RNA interference'. As a research tool, RNA interference provided, for the first time, a way to disable gene function in a rapid and flexible way.[246] No more would worm researchers have to painstakingly isolate mutants—now they could easily target specific genes by feeding worms bacteria expressing siRNAs, or just soak the worm in a solution containing such siRNAs.[241]

However, the real reason why RNA interference became such a big discovery, and would lead to the award of a Nobel Prize to Fire and Mello in 2006, was that it can be applied to many other species, including mammals.[245] In particular, the discovery that RNA interference can be applied to human cultured cells opened up exciting new possibilities for medical research, as it meant that, for the first time, human genes could be disabled in a very precise fashion. As such, it is now possible for scientists to study how 'knocking down' particular proteins affects processes in the human cell. One important use of RNA interference is in 'genome-wide screens'.[247] These use genome sequence information to create a 'library' of siRNAs covering all the genes in the genome. Cells in thousands of tiny culture dishes are then each exposed to a different siRNA, and whether this inhibits a specific cellular process has been used to identify genes involved in disorders ranging from Parkinson's disease to severe combined immune disorder, where patients lack a functioning immune system.

As well as providing new ways to diagnose human disease at the molecular level, there is much interest in using RNA interference in gene therapy. Standard gene therapy seeks to introduce a functional gene into cells where the normal gene is missing or unable to function. However, in some diseases such as cancer or types of dementia like Huntington's, it is the mutant protein that prevents the cell working normally. In such cases, selective elimination of this protein could be used to treat the disease.[248] RNA interference is so exquisitely sensitive to the sequence of the mRNA it targets, that if even a single base is different, as with the cancerous form of the RAS oncogene, then the mutant mRNA can be destroyed while leaving the normal form untouched. Unfortunately, the therapeutic potential of RNA interference has been slow to be realized.[248] While the process works extremely efficiently in human cells in culture, delivering sufficient quantities of siRNAs to the right location in a living person has proven far from easy. However, progress has been made recently in finding ways to get around this problem.[249] For instance, one approach encapsulates siRNAs in 'nanoparticles' made of a mixture of fat and protein molecules, mimicking the way in which dietary cholesterol is absorbed into cells. Using such approaches, clinical trials are currently testing the potential of RNA interference as treatments for conditions ranging from high cholesterol in the blood, to hepatitis C infection, to various types of cancer.[249]

Therapeutic usefulness aside, the discovery of RNA interference raised an important question: given the presence of RISC in cells from plants to humans, what is the normal function of this protein complex? Through addressing this question it has become clear that RNA plays a far more multi-varied, and important, role in the cell than ever suspected. The first role suggested for the RNA interference process was that it protects the cell from infection by viruses. This certainly seems to be the case in plants.[250] Viruses that infect animal cells can have genomes made of DNA like our own, as in the case of the herpes virus, or of RNA, like the influenza virus. However, most plant viruses have an RNA genome, and, maybe because of this, RNA interference is one of the primary ways in which plants combat viral infection. In invertebrate species such as worms and flies, it also plays a central role in combating infection by viruses. Interestingly, viruses have found ways to fight back against such attempts to limit them.[250] For instance, some viruses make a protein that suppresses the RNA interference machinery. Others mutate their genomes so that the siRNAs the host organism uses against them are no longer effective. However, organisms have themselves evolved ways to override these tricks, in a kind of arms race between virus and host. In mammals, however, RNA interference does not seem to play a major role in combating viral infection. This may be because we have superseded such a need with our elaborate defence system, based on the antibody proteins and other inhibitory mechanisms of our immune system.

The discovery that both plants and animals employed siRNAs as a defence against pathogens demonstrated a new importance and flexibility of roles for RNA. But an even bigger surprise was still to come, as scientists began to realize that there are many other types of non-coding RNAs that exist normally in our cells. Currently, there are four known classes of non-coding RNAs, although each class almost certainly include many subclasses. First, there are the siRNAs, which, as we've just discussed, regulate gene expression by destroying their target mRNAs. The second class are known as microRNAs, or miRNAs for short.[251] Their main mechanism of action is to prevent their target mRNA being used to make proteins by inhibiting its interaction with the ribosome. However, recent studies have shown that some miRNAs can also play stimulatory roles. So, in some circumstances, binding of a miRNA to its target mRNA enhances the latter's capacity to produce protein, while other miRNAs activate transcription of mRNAs themselves. Third, there are the piRNAs, which were first thought to be only present in the testicles and ovaries, where they act to safeguard the genomes of the eggs and sperm, an essential role given how important these are for forming the next generation.[252] However, more recently these have also been shown to play important roles in the brain. Their main action is to supress inappropriate 'mobility' of DNA elements in the genome, in ways we'll soon be exploring. The fourth class are the long non-coding RNAs, or lncRNAs.[253] These are defined mainly by length, all being over two hundred bases long, in contrast to the other three classes which are typically much smaller, at around twenty bases. These RNAs have various ways of regulating gene expression, but one particularly important role is to bring different parts of the genome together to form a complex 3D network of functionality, in ways we'll look at shortly. Between them, these different classes of non-coding RNAs act to regulate normal gene expression at practically every stage of this process. Since we are going to encounter non-coding RNAs throughout the remainder of this book, for now I just want to focus on one class—the miRNAs—in order to demonstrate the diversity of physiological processes that even one class of non-coding RNAs is involved in regulating.

It now seems that as many as half of all human genes are regulated by miRNAs.[254] Such miRNAs are related in sequence to the mRNAs of the genes they regulate, but typically can control multiple mRNA targets. Although miRNAs regulate a range of processes from embryo development through to adulthood, they have particularly pronounced roles in certain areas.[251] One such role is in stem cells: these cells are found throughout the body, but are particularly active in tissues that need continual replenishment like skin, blood, the gut lining, and the male gonads. Stem cells can divide indefinitely, but also generate all the cell types of the organ or tissue in which they are found. Some miRNAs are expressed in particular types of stem cells. By manipulating the mouse genome to prevent the expression of these miRNAs, it has been possible to understand what functions they play in these cell types. Such manipulation of the genome makes use of another type of stem cells, so-called 'embryonic' stem cells, which have the potential to turn into any cell in the body. By modifying the genomes of such cells and then using them to create a mouse embryo, it is possible to generate a whole mouse with the gene modification.[255] While this approach has been mainly used to 'knock out' protein-coding genes in the mouse, in order to study their role in health and disease in a living mammal, it can be applied to any functional genomic element.

This approach has shown that miRNAs play key roles in many processes mediated by stem cells. One dramatic effect of blocking miRNA action is on the developing sperm.[251,256] A typical man produces about a thousand or so sperm in the time taken for one breath. What fuels this prodigious production process are the testicular stem cells. But when a specific miRNA only found in the testicles was knocked out, sperm production ground to an abrupt halt. Since some men are infertile because they fail to produce any sperm, there is now interest in whether this may sometimes be due to an miRNA defect. Conversely, a male contraceptive drug could be designed to specifically target the testicular miRNA; it would have to be very specific in its action though, for miRNAs have been shown to play other vital roles in the body and there could be severe consequences if these were blocked.

For instance, miRNAs play important roles in formation of blood cells. The complex nature of such regulation was shown by the fact that knocking out different miRNAs in mice upset the balance of the different types of blood cells in very specific ways.[251] So, while knocking out one miRNA led to the loss of certain white blood cells and a serious loss of immunity as a consequence, knocking out another depleted the red blood cells that carry oxygen in the blood, with resulting anaemia. In line with such a role, recent studies have shown that abnormal expression of miRNAs in humans is associated with some types of leukaemia— cancers of the blood.[257] Skin, hair, and the gut lining are all rapidly dividing tissues that rely on continual renewal by stem cells, so maybe it is not surprising that these too are badly affected when the miRNAs in such tissues are disabled.[251] Another important site of action for miRNAs is in the nervous system and brain.[251] So, inhibition of miRNA action in astroglial cells, which act as a support network for nerve cells, resulted in brain dysfunction and seizures. Curiously though, inhibition of miRNAs in the forebrain resulted in an initial enhancement of learning and memory until, eventually, nerve degeneration set in. This provocative finding raises the question whether interfering with miRNAs in humans could promote learning in some circumstances. Interestingly, an important effect of brain miRNAs is on transcription factor CREB, which, as we saw in Chapter 3, plays a key role in learning and memory.[251] Other miRNAs regulate the heart and circulation.

So, starting from a failed attempt to turn a petunia purple, the discovery of RNA interference has revealed a whole new network of gene regulation mediated by RNAs and operating in parallel to the more established one of protein regulatory factors. However, another surprising discovery has emerged from the study of miRNAs: their point of origin. Studies have revealed that a surprising 60 per cent of miRNAs turn out to be recycled introns, with the remainder being generated from the regions between genes.[258] Yet these were the parts of the genome formerly viewed as junk. Does this mean we need a reconsideration of this question? This is an issue we will discuss in Chapter 6, in particular with regard to the ENCODE project, with its controversial findings that pose a challenge to the current consensus about how genes work, provoking both excitement but also deep disagreement.

6

IT'S A JUNGLE IN THERE!

'It's likely that 80 percent will go to 100 percent. We don't really have any large chunks of redundant DNA. This metaphor of junk isn't that useful.'

Ewan Birney

'Just because a piece of DNA has biological activity does not mean it has an important function in a cell. Most of the human genome is devoid of function and these people are wrong to say otherwise.'   *Dan Graur*

Metaphors have been central to biology as far back as Aristotle. In the seventeenth century, French scientist René Descartes compared animals to machines, and his English contemporary William Harvey described the heart as a pump. In the early twentieth century, the biochemist's view of the cell as a factory dominated, with structural proteins its bricks and mortar, and enzymes the machines that carry out the manufacturing process. As awareness grew of the commanding role of genes in cellular life, so did descriptions of the nucleus as the manager's office controlling activities on the factory floor of the cytoplasm. And, as we saw in Chapter 2, following the discovery of the DNA double helix the idea of life as a digital code became dominant. With the four nucleotide bases acting as the letters for this code, perhaps it was not surprising that the genome has become known as the book of life, although this description was clearly too old-fashioned for Walter Gilbert. He developed his own version of DNA sequencing alongside Sanger in the 1970s, sharing a Nobel Prize with the latter in 1980, and, from the late 1980s, became a central proponent of the human genome project. As part of his fundraising efforts at the time, he ended his seminars by holding up a glittering CD to the audience and declaring that 'soon I will be able to say "here is a human being; it's me"'.[259] Soon molecular biologists across the world were using this high-tech image to sell the idea of a digitalized approach to biomedical science. However, at the completion of the project over a decade later in 2003, digital downloads were already making the CD metaphor look almost as dated as that of the book it replaced. A more fundamental problem with the metaphor had arisen though, for an e-book, like its paper form, is generally still expected to have a clear structure, with an introduction, conclusion, and a narrative sandwiched in the middle.

Yet an unexpected message of the human genome project was that the number of genes was much less than expected, and the proportion of non-coding DNA substantially greater than imagined. If the genome were a book, it seemed hard to escape the fact that the vast majority of it was complete gibberish. Not only that but the genes themselves were far from transparent, revealing little by DNA code alone about the cellular processes they mediated, the cell types and tissues they were expressed within, and the signalling pathways that regulated them, apart from what was already known from previous studies. Reflecting such concerns, the same year as the genome project was completed, a major new initiative was launched. Named ENCODE, for ENCylopedia of DNA Elements, its stated aim was to characterize all the 'functional' elements in the genome.[260] This seemed an acknowledgement that one book of life wasn't enough, and that while the human genome project might be compared to a dictionary, albeit a minimal one that merely listed the different words in a language, so an encyclopaedia was needed to provide detailed information about each gene and its relationship to the other genes in the genome. This was because, as Sydney Brenner noted about the genome project, 'getting the sequence will be the easy part as only technical issues are involved. The hard part will be finding out what it means, because this poses intellectual problems of how to understand the participation of the genes in the functions of living cells.'[261]

Importantly, ENCODE's scope was not to be limited to the protein-coding genes but would instead extend across the whole genome.[260] In part this reflected a growing recognition, stimulated by some of the discoveries we discussed in Chapters 4 and 5, that maybe some of the 'junk' in the genome was not as useless as had been supposed. In recognition of this fact, and showing that biologists are happy to borrow metaphors from whatever source they can, a new phrase increasingly used to describe the non-coding DNA, was that it represented the genome's 'dark matter'.[262] This term is more commonly used by cosmologists to refer to the 85 per cent of the universe that is invisible to the naked eye, and indeed to any telescope, but is, nevertheless, presumed to exist to account for discrepancies between the universe's observed gravitational behaviour and its mass, implying that there must be much more mass than we can observe.[263] Given the difficulties in directly confirming the existence of this cosmological dark matter, or indeed telling us anything about its properties (although a recent study may have detected it in the Sun[264]), this might not seem a very helpful metaphor. But at least it conveyed the feeling that non-coding DNA could have an important function and was therefore worth investigating seriously.

ENCODE's initial focus was relatively cautious—a pilot project beginning in 2003 that surveyed only 1 per cent of the human genome. Its findings, published in 2007, were tantalizing, since they suggested that non-coding DNA was far more active than supposed.[265] This conclusion was sufficient to secure further major funding to survey the remaining 99 per cent of the genome. Involving 442 scientists from 32 institutions, and costing $288 million, the ENCODE project was clearly an example of 'big science'.[260] An important issue was how genome function would be assessed. Unlike the original genome project—a straightforward, if daunting, matter of reading the 3 billion bases in our genome using a single method—diverse approaches would be required to study the genome as a living, functioning entity rather than just a series of letters (see Figure 19). One method involves identifying all the places in the genome where transcription factors bind to the DNA. As we saw in Chapter 3, such factors come in different shapes and sizes and relay information to the genes they control from diverse cellular signals. Historically, these factors had been studied one by one, with no sense of how their activity looked at a global level. However, one consequence of the genome project was the development of techniques that made it possible to do this. One such approach uses antibodies generated against transcription factors to purify such factors while they are still attached to their DNA target.[260] The latter can then be sequenced, creating a map showing where each regulatory factor binds in the genome.

Another approach measures differences in accessibility of different parts of the genome to identify regions of activity. Because control elements of active genes are less tightly bound to histones, which, as we saw in Chapter 3, wrap around the DNA but, when acetylated, do so much less tightly, they are both more accessible to transcription factors and also to enzymes called DNAases that cut unprotected DNA into fragments. By identifying parts of the genome that are susceptible to

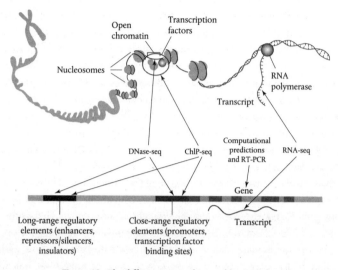

Figure 19. The different approaches used by ENCODE

being cut by such DNAases, a map can be constructed of those regions that are most accessible to transcription factors.[260] In fact, recent studies have shown that acetylation is merely one of a bewildering variety of chemical changes that affect histones, others being methylation and phosphorylation. The sheer number of these changes, each differently affecting gene expression, and the fact that such changes are highly responsive to incoming cellular signals, has given rise to the idea of a 'histone code' that operates parallel to the DNA code.[266] Importantly, whereas initially changes to histones were only thought to affect their association with DNA, more recent studies show that these can also act as a signal to recruit other proteins that regulate gene expression. Another important approach used by ENCODE involved cataloguing these histone modifications in the genome.

A third approach ENCODE used to investigate genome activity was surveying chemical changes in the DNA.[260] Studies in the 1980s first showed that addition of methyl (chemical structure -CH_3) groups to C nucleotides within genes and their regulatory regions profoundly affects their expression. As such, identification of methylated C nucleotides across the genome is another indicator of gene activity.[266] Finally, ENCODE studied activity in the genome by cataloguing its

RNA output.[260] Previously, studies of RNA expression had focused on mRNA. However, with the discovery of the diverse types of non-coding RNAs that we discussed in Chapter 5, the focus shifted to encompass these too.[267] In addition to employing multiple techniques, another important feature of ENCODE, compared to the genome project, was its study of many different cell types.[260] For while all cells have the same basic genome—bar exceptions like red blood cells with no nucleus, and eggs and sperm that have only half the DNA content of a normal cell—the ways in which genes can be turned on or off is as great as the number of different cell types in the body. As such, ENCODE studied genome activity in 147 different cell types.

Undoubtedly, the most surprising aspect of ENCODE was its discovery that, far from being inactive compared to the protein-coding genes, the other 98 per cent of the genome was humming with life, prompting Ewan Birney, the charismatic spokesperson for the project based at the Sanger Centre near Cambridge, England, to say 'it's like a jungle in there. It's full of things doing stuff.'[268] Initial claims from ENCODE leaders were that as much as 80 per cent of the human genome had a biochemical function. In fact, Birney went beyond even this, saying, 'it's likely that 80 percent will go to 100 percent'.[268] Echoing this, Thomas Gingeras of Cold Spring Harbor Laboratory, based at Long Island in the USA, said 'almost every nucleotide is associated with a function of some sort or another, and we now know where they are, what binds to them, what their associations are, and more'.[268] This conclusion was so radically different from the idea that most of the genome was 'junk', that John Stamatoyannopoulos of the University of Washington, predicted that the findings would 'change the way a lot of concepts are written about and presented in textbooks'.[269]

Based on studies that localized binding sites for regulatory proteins across the genome to 3.9 million different regions, Birney estimated that over 4 million 'switches' were scattered around the genome, devoted to controlling the activity of the genes—a startling imbalance given there are only just over 22,000 of the latter.[268] But it wasn't only gene switches that were detected in abundance. The study also found that 80 per cent of the genome was generating RNA transcripts.[260] In line with these transcripts having functional importance, many were found only in specific cellular compartments, indicating that they have fixed addresses where they operate. Surely there could hardly be a greater divergence from Crick's central dogma than this demonstration that RNAs were produced in far greater numbers across the genome than could be expected if they were simply intermediaries between DNA and protein. Indeed, some ENCODE researchers argued that the basic unit of inheritance should now be considered as the transcript. So Stamatoyannopoulos claimed that 'the project has played an important role in changing our concept of the gene'.[270]

Another exciting development was ENCODE's claim to have provided an explanation for a problem that had confounded researchers ever since attempts began to identify the links between common diseases and specific parts of the genome. Such 'genome-wide association studies' work on the principle that, by taking a sufficient number of people—say half a million individuals who, being human, will suffer from various common diseases—and surveying common DNA sequence variants across the human genome, it should be possible to show which variants are associated with a particular disorder.[271] This, then, should identify the important genes, which, when defective, lead to someone succumbing to such a disorder. Since the completion of the genome project, many such studies have been carried out for issues ranging from cardiovascular disease, diabetes, and stroke, through to problems of the mind like schizophrenia and bipolar disorder.[272]

Some important DNA sequence variants have been linked to human disorders by this strategy. But a major problem in interpreting the significance of such findings was that 90 per cent of the DNA variants uncovered in this way were nowhere near any protein-coding genes.[271] However, many of these previous links with disease overlapped with regions of genomic activity identified by ENCODE. Importantly, regions previously implicated in a particular disease were active in precisely the cell type in which this disease occurs.[273] For instance, Birney pointed to the example of Crohn's disease, 'a pretty awful gut disease where the body attacks its own cells. There are a hundred places in the genome where we know a genetic change increases the risk of getting Crohn's. Many of those overlap with a switch identified by ENCODE.'[274] Particularly relevant, given that the body's immune system turns upon itself in Crohn's disease, was the fact that such genomic regions were active in T lymphocytes, which play a central role in mediating immunity. Such insights came from ENCODE surveying many different cell types. This analysis showed that although 4 million gene switches exist in total, which ones are active depends on cell type. This discovery emphasized the importance of cellular environment for gene expression, on a much greater scale than had been realized.[273]

Another surprise of the ENCODE project was how little conservation of genetic switches it detected between humans and other mammals. The project had been specifically carried out in parallel with a similar survey of the mouse genome to study this question. Although the similarity between mice and humans at the protein level is 85 per cent, ENCODE's comparison of regulatory regions in the human genome compared to that of the mouse showed a similarity of only 50 per cent.[275] Could this mean that while the proteins of the two species were largely alike, the factors controlling them were widely different? To investigate whether regions of the human genome unique to our species have important functional roles, comparisons of different human individuals were carried out by ENCODE researchers as part of the so-called '1000 Genomes Project'. This indicated that as much as 4 per cent of our genome that is not shared with other mammals is preserved amongst different people, suggesting it is newly under the influence of natural selection and, therefore, important.[275]

The ENCODE findings were widely reported both in glowing editorials in top science journals like *Nature* and *Science* but also across the media. Overall, the feeling was that here was a major new advance in our understanding of how the genome worked, with implications both for our understanding of our biology, and for the diagnosis and treatment of disease. Yet not everyone was so enamoured. Some months after the findings were published, an article, headed by Dan Graur of Houston University, appeared in the journal *Genome Biology and Evolution*,[276] attacking the findings in a vitriolic tone not normally associated with scientific debate, or at least not in the pages of an academic journal. According to the article, the claims of ENCODE were 'absurd', its statistics 'horrible', and it was 'the work of people who know nothing about evolutionary biology'. And, in response to claims by John Stamatoyannopoulos that the findings would necessitate a 're-writing of the textbooks', the article countered by saying that 'the textbooks dealing with marketing, mass-media hype, and public relations may well have to be rewritten'.[277] In a subsequent interview, Graur reiterated these points and added more caustic ones, claiming that 'this is not the work of scientists. This is the work of a group of badly trained technicians.'[278] And just in case his disdain for the project hadn't been fully appreciated, Graur also began showing a slide at the end of his presentations on the topic of a photograph of dollar bills taped together in the shape of a toilet paper roll—his view of what ENCODE had achieved with the $288 million spent on the project.[279]

So what was the motivation for such attacks? One claim made by Graur and his co-authors was that ENCODE researchers had confused biochemical activity with function. So just because huge areas of the genome are peppered with binding sites for regulatory proteins, modified histones, and evidence of large-scale RNA production outside of the protein-coding genes, this does not necessarily mean all are functional.[276] Rather, such activity may merely be genomic 'noise'. After all, if much of the genome can be junk, is it not possible that it may also generate junk activity? A central criticism raised by Graur and his co-authors was that a huge discrepancy existed between the large proportion of the human genome ENCODE leaders claimed was functional, compared to that predicted to be under the pressure of natural selection. As we saw in Chapter 1, variation within a species is believed to be ultimately due to changes in the genome caused by mutation, while the continued presence of such variants is linked to whether they have the characteristics necessary for survival in a particular environment, or are neutral enough in their effects that there is no pressure to eliminate them. But how can the influences of natural selection be gauged across the genome as a whole? Traditionally, the approach used is to compare the complete sequences of the genomes of different species, nucleotide by nucleotide. When used to compare the protein-coding portion of the genome, humans are seen to be 99 per cent similar to a chimpanzee,[280] 85 per cent similar to a mouse,[281] and, confirming the link between all life forms on the planet, even 50 per cent similar to a banana.[282]

But if the whole human genome is compared to that of the mouse, the similarity is far less, only around 5 per cent. And although a recent study comparing the genomes of different human individuals indicates a further amount of similarity that we all share, this only appears to be a further 4 per cent.[283] But, according to Graur and his co-authors, this made a mockery of the ENCODE claim that 80 per cent of our genomes are functional.[276] In particular, they accused ENCODE leaders of ignoring the role of 'purifying selection'. This is the process whereby natural selection prevents changes to DNA sequences in the genome because of their usefulness to the organism in which they occur. Without this selective pressure, there is nothing to stop mutations occurring and changing a particular sequence. But if only 9 per cent of the genome is 'conserved' in this way, then by definition only it can be functional, otherwise the conservation of sequence would be much greater. And, according to Graur and his co-authors, that means that the ENCODE claims of 80 per cent of the genome being functional are 'absurd', because they 'assume that no deleterious mutations can ever occur in the regions they have deemed to be functional'. Indeed, such claims are akin to believing that 'a television set left on and unattended will still be in working condition after a million years because no natural events, such as rust, erosion, static electricity, and earthquakes can affect it'.[284]

Although the concerns raised by Graur and his co-authors were primarily directed at the underlying science of the ENCODE project, a more veiled concern of the article was the implications of the project's claims for the debate between creationists and evolutionists. We saw in Chapter 4 how the discovery of what appeared to be useless junk in the genome was used as evidence against the idea that a supernatural being designed life, including our own species. But if 80 or even 100 per cent actually turns out to be important, this might be seen to undermine the case for natural selection and let 'intelligent design' in through the back door. Or at least that seemed to be a concern of Graur and his co-authors, given that the title of their article was 'Function in the human genome according to the evolution-free gospel of ENCODE', and elsewhere they said that 'the only people that should be afraid of junk DNA are those claiming that natural processes are insufficient to explain life and that evolutionary theory should be supplemented or supplanted by an intelligent designer'.[285]

Finally, Graur and his co-authors used the opportunity to take a swipe at 'big science'. One consequence of the global recession is that obtaining the government grants that fund most university professors' research is becoming increasingly difficult. As such, there is increasing disquiet amongst some researchers that 'big science' projects like ENCODE are taking the bulk of available money. No matter that a single military fighter jet, at $350 million, can cost more than the combined ENCODE funding,[286] the accusation that the project was not just expensive but incompetent has achieved some resonance with researchers conducting the sort of 'small science' that has been the traditional norm, but who are now faced with a decreasing central pot of money.[284] Indeed, Graur reiterated this point about ENCODE in a recent interview, saying that 'when the average grant size in the biomedical sciences has been halved compared to 10 years ago, this is a scandal. If you pour $288 million into one project, you do not fund 500 other projects. You kill the careers of young scientists. They are reduced to becoming technicians.'[287]

Faced with such a strongly worded attack, the response from ENCODE researchers might have been expected to be similarly robust, or at least to address the key points raised by Graur and his colleagues. However, in an interview with *The Guardian* newspaper following the *Genome Biology and Evolution* paper, Ewan Birney would only say that 'the nature of the attacks against us is quite unfair and uncalled-for'. Birney also took part in an interview at the BBC with Chris Ponting of Oxford University, who had also criticized the ENCODE claims, albeit in more measured tones than Graur and his colleagues. Questioned by Ponting as to how much of the genome was 'vital for life', with the suggestion that this might be between 10 and 20 per cent, Birney seemed to agree with this suggestion, which left the basis for the previously claimed much higher figure of functionality unclear. Finally, asked by the BBC interviewer whether the ENCODE leaders had 'let the story get a bit away' from them, Birney's response was only that 'hindsight being such a cruel thing, [this] makes me think about what I could have done to minimize this kind of rather heated debate'.[288]

All of this was rather disappointing for anyone hoping to see a more robust defence of the original ENCODE claims, or an answer to the charge that such claims had been misplaced, either through incompetence or a desire to make a big splash with the media. However, others were prepared to be more vigorous in their defence. In particular, John Mattick and Marcel Dinger, of the Garvan Institute of Medical Research in Sydney, wrote an article for the *HUGO Journal*, official journal of the Human Genome Organisation, entitled 'The extent of functionality in the human genome'.[289] In this they responded in detail to the article by Graur and his co-authors. In response to the accusation that the apparent lack of sequence conservation of 90 per cent of the genome means that it has no function, Mattick and Dinger argued that regulatory elements and non-coding RNAs are much more relaxed in their link between structure and function, and therefore much harder to detect by standard measures of conservation. This could mean that 'conservation is relative', depending on the type of genomic structure being analysed.[289] Secondly, against the idea that the huge numbers of RNA transcripts produced by the genome are mainly random noise, Mattick and Dinger pointed to the fact that ENCODE had confirmed many preceding studies demonstrating that transcripts were produced in 'cell-type specific patterns' and showed 'dynamic regulation in embryo development, tissue differentiation, and disease', suggesting they play an important role in such processes.[289] The third point tackled was the insinuation that claiming a large degree of function across the genome necessarily plays into the hands of those who want to reject the idea of biological evolution driven by natural selection. In answer to this, Mattick and Dinger argued that a high degree of functionality in the genome was 'entirely consistent with the tenets of evolution by natural selection', albeit along different lines to those who held to a view of the genome that was primarily focused on protein-coding genes.[290]

That Mattick was prepared to make such bold claims was no surprise to those familiar with his previous views on this topic. In fact, well before the ENCODE findings were published in 2012, he had been making the case that the genomes of multicellular organisms were far more complex in their mechanism of operation than had been imagined. So, in an article published in 2007 following the results of the initial ENCODE pilot project, Mattick said 'it is also now clear that the majority of the mammalian genome is expressed and that many mammalian genes are accompanied by extensive regulatory regions'.[291] In arguing for such a viewpoint, Mattick has suggested that the original concept of the genome as a digital code did not go far enough.[292] According to him, when Watson and Crick came up with their revolutionary proposal in the 1950s, one problem was that their views were coloured by the society of the time, this being a world of analogue devices like vinyl records and slide rules. As such, they conceived of the gene primarily as a recipe—albeit using a digital code—for analogue devices, the proteins. This led to the idea of a simple one-way transfer of information from DNA to proteins, which led to Crick's central dogma.

Yet now that we live in the digital age, we recognize that digital information can be highly multi-layered. And, according to Mattick, biologists are only just beginning to recognize that such multi-layered information is characteristic of our own genomes, with different forms of non-coding RNAs playing multiple roles in the different layers. In line with this idea, in a recent article in *Nature Reviews in Genetics*, entitled 'The rise of regulatory RNA', Mattick has claimed that 'RNA is the computational engine of cell biology, developmental biology, brain function and perhaps even evolution itself. The complexity and interconnectedness of these systems should not be cause for concern but rather the motivation for exploring the vast unknown universe of RNA regulation, without which we will not understand biology.'[293] A proponent of a similar viewpoint is Evelyn Fox Keller of the Massachusetts Institute of Technology, who believes recent 'genomic science has changed the very meaning of the term, turning the genome into an entity far richer, more complex, and more powerful—simultaneously both more and less—than the pre-genomic genome, in ways that require us to rework our understanding of the relation between genes, genomes and genetics'.[294]

So who is right? Has ENCODE opened a new chapter in our understanding of the genome and how it works, or are the conclusions of the project's leaders flawed and misleading? To address this issue, it's time we began to dig deeper into those proposed multiple layers of the genome, like miners trying to find the richest seams, in order to gain further insights into what these layers are, and their relationship to each other. In doing so, we will not only look further at the role of the various types of non-coding RNAs, but also at the histone proteins that wrap around DNA. In addition, we will examine the chemical modifications that alter these proteins and the DNA itself. We will also study how such modifications are affected by changes in the environment of the cell and organism, and whether the genome as an entity is as stable as we have been led to believe. But before we do any of that, it's time to consider whether the original idea of the genome as a primarily linear object stretched out along a chromosome still holds, or whether, also like a mine, it is better understood in 3D. So if you are ready, let's get digging!

7

THE GENOME IN 3D

'Natural DNA is a tractless coil, like an unwound and tangled audiotape on the floor of the car in the dark.'   *Kary Mullis*

'The genome is like a panel of light switches in a room full of lights. These switches can be located far from the genes they regulate in the one-dimensional genome sequence but in three dimensions, the chromosome is folded so that they physically touch.'   *Job Dekker*

It's time to give the genome some physicality and some shape. A common view of chromosomes is of linear strands of DNA upon which the genes are dotted like beads on a string. In a very obvious sense chromosomes are linear entities, since each is an unbroken chain of bases, ranging from a quarter of a billion for the largest human chromosome 1, to 50 million for the smallest, chromosome 22.[295] Proteins are also linear molecules, albeit magnitudes smaller than even the smallest chromosome, but it has long been recognized that it is the 3D ordering of the amino acid chain that gives each specific protein its characteristic and unique properties. Now, however, there is a growing recognition that chromosomes too are complex 3D entities, so much so that ENCODE researcher Job Dekker of the University of Massachusetts Medical School recently said 'nothing in the genome makes sense, except in 3D',[296] paraphrasing a previous comment from renowned evolutionary biologist Theodosius Dobzhansky, who once said 'nothing in biology makes sense except in the light of evolution'.[297] So how did Dekker come to this conclusion and how exactly does a 3D genome function?

One factor that helped shape our image of chromosomes as solely linear entities was the genetic maps that scientists from Morgan and Sturtevant onwards began to construct, which reinforced the notion of such linearity. And while it is ethically impossible to subject humans to massive doses of X-rays in order to stimulate mutations, or carry out mating experiments to study their transmission across generations, enough information has accumulated from investigations of families in which certain characteristics or diseases can be followed, to allow construction of such maps for our own species. So even before the genome project, detailed maps already existed of the relative positions of many human genes on each chromosome.[298]

The idea of chromosomes as linear entities was also reinforced by the shape chromosomes assume during the phase of cell division known as mitosis, when they become the tiny threads that Walther Flemming first observed, as we saw in Chapter 1. Chromosomes assume such a condensed linear form to ensure the two chromosome pairs segregate to each daughter cell in a tangle-free fashion.[299] Yet apart from this brief period in a cell's life, chromosomes normally assume a much looser form, with their DNA chains distributed across the 3D space of the cell nucleus. Not that this distribution is random; instead, it's been known for some years that DNA is densely packed. Indeed, it would be hard to imagine how the immensely long genome could be accommodated within the small nuclear space otherwise.[300] If all the DNA in a human cell were laid end to end, it would stretch for two metres. Yet the nucleus is only ten microns—less than one hundred thousandth this size—so how is this packaging problem solved?

The discovery of histones went a long way to providing an answer. By coiling the DNA into nucleosomes, the interaction with histones provides the first level of genomic packaging, but also makes possible a series of further levels of organization counterposed on top of each other like a set of Russian dolls.[301] Another important contributor to genomic packaging is the nuclear lamina, a fibrous network concentrated near the periphery of the nucleus that helps to organize the nuclear pores, holes in the nuclear membrane that allow molecules to travel between the nucleus and the rest of the cell. Recent studies have shown that the genomic regions associated with the nuclear lamina are those with low activity, which explains why early studies of stained cells first showed that the most condensed chromatin—the complex mixture of DNA, histones, and other proteins that genomes are composed of—is particularly concentrated around the periphery of the nucleus.[302]

The one type of human cell in which DNA is not primarily associated with histones is the sperm. We saw in Chapter 2 how Friedrich Miescher first isolated and characterized DNA from white blood cells in the pus from surgical bandages while a young scientist working in Tübingen, Germany. Later, now based in his home city of Basel, Switzerland, he found another, slightly more salubrious source of DNA in salmon sperm, which he obtained in large quantities from the river Rhine near his laboratory. However, although his colleague Albrecht Kossel had shown that normally DNA is associated with histones, Miescher found that this is not true of the sperm: instead, its DNA is associated with another protein which he called protamine.[303] Histones and DNA are attracted to each other because of their respective basic and acidic chemical properties. Protamines are even more basic than histones and consequently have a greater affinity for DNA. Because of this, the sperm genome is packaged in an almost crystalline fashion. This allows the DNA to fit into the highly streamlined sperm head, an essential feature for a cell that must swim a great relative distance before it gets to its target, the egg. And just as Olympic swimmers shave or squeeze into high-tech supersuits to further streamline themselves, a recent study by scientists at the European Molecular Biology Laboratory in Grenoble showed that sperm streamlining is boosted by a protein called BRDT.[304]

The study showed that BRDT drives the replacement of histones by protamines by adding acetyl tags to the former. According to Saadi Khochbin, who led the study, 'in sperm, just before the DNA starts to hypercompact, these tags are added throughout the chromatin in a huge wave. If BRDT is absent, the extra compaction doesn't take place, and the sperm head would be less streamlined.'[305] Demonstrating the importance of this process for normal sperm function, male mice lacking BRDT are infertile. As well as aiding streamlining, such tight packaging helps to protect the sperm DNA from the potentially harmful bodily chemicals to which it is exposed during its journey through the female reproductive passage. However, recent studies have shown that a significant proportion of human sperm DNA—up to 15 per cent as opposed to only 1 per cent in mouse sperm—is associated with histones, with the genes first activated during embryo development being packaged in this way, in line with the idea that this looser association allows more rapid access of this part of the genome to transcription factors.[306] That the packaging of any particular region of the genome might affect its activity was first suggested many years ago by microscopy studies of cells exposed to different kinds of chemical stains, which revealed that chromatin exists in two main forms—a dark-staining tightly packed version called heterochromatin and a lighter-staining, looser form called euchromatin.[307] In addition, recent studies indicate that, far from being a static structure, the 3D genome in the living cell is also highly dynamic, with structural changes in the chromatin being intimately linked to gene activity. Such recognition has come from experimental approaches that visualize the position of specific genes within the nucleus.

One such approach is called fluorescence *in situ* hybridization, or FISH for short. The fluorescently labelled DNA probes used in FISH are complementary to specific sequences in the genome and can be used to specifically identify the presence of a gene within a chromosome. 2D FISH has become an important clinical tool in the diagnosis of genetic abnormalities, particularly those associated with cancer.[308] In this approach, the probe is used to identify whether specific oncogenes are amplified in the genome. Normally, a gene is recognized as two spots of fluorescence, since there are two gene copies in a typical cell. However, in cancer, the cellular machinery that duplicates the genome at each cell division often breaks down, with some parts of the DNA being replicated more than once. And it's not a coincidence that the genome regions that tend to become amplified are those containing genes involved in cell growth and other processes that are subverted in cancer. Darwin and Wallace's theory of natural selection as the driving force of evolution views individual organisms in a species as subject to the 'survival of the fittest'. However, this principle is also central to cancer, as cells that overcome the normal limits governing cell growth and other forces limiting tumours, are selected for their superior qualities in this regard.[309] Through amplification of genomic regions with a high concentration of oncogenes, a cancerous cell can gain an advantage compared to other cells in the tumour. RAS is a particularly important oncogene because of the central role its non-mutated form plays in normal cell growth; it is amplified in up to 30 per cent of human cancers. 2D FISH can be used diagnostically to see whether this gene has become amplified in a cancer cell, since if this is the case, instead of two fluorescent spots, many more will be observed.

A further modification of 2D FISH is called chromosome painting.[310] In this approach the fluorescent probe targets DNA sequences across a whole chromosome. By using a different colour fluorescence for each chromosome, this technique makes it possible to easily identify all the different chromosomes in a cell. Previously, this meant laboriously comparing their different sizes and staining patterns with chemical dyes—such dyes give a pattern of bands unique to each chromosome but which can only be distinguished by someone trained to recognize the subtle differences. Chromosome painting is used to identify people with genetic diseases involving chromosome 'translocations', where parts of one chromosome break off and become incorporated into a different chromosome during cell division.[310] By disrupting normal gene expression, such translocations can lead to disease, including cancer.

However, as well as being used to study condensed chromosomes, FISH has recently been used to study the location of genes in the uncondensed DNA that fills the 3D space of the nucleus in the so-called 'interphase' period between cell divisions.[311] Visualizing fluorescence in this 3D space is carried out using so-called 'confocal' microscopy. Typically, this uses a laser beam to scan across a 3D section and pick out a spot of fluorescence in high resolution. Combining this technique with FISH has made it possible to identify the precise position of a gene within the nucleus.[312] Such studies have shown that there are 'active' and 'inactive' regions of the nucleus, with genes that are switched on in the former, and genes that are switched off in the latter. Remarkably, when a gene's activity changes because of a cellular stimulus, its position in the nucleus also changes. Such changes occur, for instance, during stem cell differentiation: the process by which an unspecialized, rapidly dividing cell gives rise to a specialized cell type. 3D FISH showed that genes that keep the stem cell in its unspecialized state move into an inactive region upon differentiation, whereas genes that give the differentiated cell its specialized character move into an active region.[301] This shows that the nuclear space is a far more dynamic entity than previously thought.

Also revealed by recent studies is the complexity of the interactions between different genomic regions that occur within the 3D nuclear space. We saw in Chapter 4 how the discovery of enhancers was initially baffling since they can operate at a great distance from the genes they regulate. However, subsequent studies showed that enhancers, and the transcription factors bound to them, loop around to the gene promoter, and in this way can influence its expression. As important as these studies were, they were very much focused on the individual gene. However, over the last decade, new approaches that allow scientists to 'capture' interactions between different parts of the genome are revolutionizing our understanding of how these parts fit together. Such approaches use chemicals to cross-link the proteins that bind DNA, and then advanced sequencing technology to identify the DNA sequences to which such proteins are attached.[313] Importantly, this analysis is done on a 'global' scale so that all the interactions occurring in the genome can be studied at once. By carrying out such analysis on a variety of cell types following physiological stimulation, it is becoming possible to build up a picture of how such interactions change during different cellular events.

Although the approaches discussed so far have greatly enhanced our understanding of how the genome operates in the 3D nuclear space, one significant limitation is that none involve study of the living cell. So 3D FISH is carried out on cells fixed with formaldehyde and then incubated with a fluorescent RNA probe. Similarly, methods that capture interactions between different genomic regions use cross-linking chemicals that kill the cell but freeze the molecular interactions within, and then fragment the DNA and associated protein. Although comparisons can be made between cells at different stages of differentiation or physiological stimulation, this is still a static picture. However, recently it has become possible to study genome interactions in a living cell in real time, through the use of fluorescent proteins first discovered in jellyfish, but which have since revolutionized the study of cellular processes by making it possible to fluorescently 'tag' molecules and follow their movement and activity.[314]

When Japanese biologist Osamu Shimomura, working at the Woods Hole Marine Biological Laboratory near Cape Cod, began studying why certain jellyfish were a striking green fluorescent colour, his main incentive was pure curiosity. But his discovery that a single protein—green fluorescent protein or GFP for short—was responsible for this property and the subsequent isolation of the gene coding for this protein, raised the possibility that GFP could be used to make the protein products of other genes 'visible' by genetically fusing the GFP gene to them.[314, 315] In fact, it was another scientist at Woods Hole, Douglas Prasher, who first cloned the GFP gene and suggested that it could be used as a fluorescent tag. However, Prasher was unable to obtain research funding to push his idea forward and instead two other scientists, Martin Chalfie and Roger Tsien, developed GFP in this way. Tsien had already found fame creating small molecules that fluoresce when they come into contact with cellular messengers, such as the calcium ions and cAMP that we discussed in Chapter 3.[316] This made it possible, for the first time, to 'visualize' changes in the concentrations of such messengers in the cell. Now, by mutating GFP in various ways, Tsien created a range of differently coloured fluorescent proteins, which, showing a characteristic humorous streak, he named after fruits.[314] So, in research papers it is now common to read about proteins tagged with banana, plum, tomato, grape, and so on. Quirky names aside, the generation of differently coloured proteins that fluoresce at different wavelengths of light has made it possible to study the localization of two or more proteins in the cell at the same time, by giving them differently coloured tags.[317]

The importance of GFP was recognized by the award of a Nobel Prize to Shimomura, Chalfie, and Tsien in 2008. In his acceptance speech, Tsien drew attention to the fact that 'aspects of our work were fragile results of lucky circumstances' and how 'funding was difficult at times to obtain for basic research on obscure organisms like the jellyfish that was the source of GFP', adding that he hoped the award of the prize would reinforce 'recognition of the importance of basic science as the foundation for practical benefits to our health and economies'.[318] As a poignant illustration of this issue, during the Nobel announcements it emerged that Doug Prasher, who had originally cloned GFP, had left science, having failed to obtain funding for his research, and was working as a courtesy shuttle driver.[319] Subsequently, Tsien, who had always championed Prasher's input, not only paid for Prasher to attend the Nobel celebrations in Stockholm, but also offered him a job as a senior scientist in his laboratory. In his speech Tsien also noted the potential effects of environmental destruction, saying how 'over the last ten years, observed numbers of jellyfish in their Pacific Northwest habitat have declined by over a thousandfold ... what other potential scientific breakthroughs may never happen because of man-made pollution and global warming?'[318]

Use of fluorescent tags such as GFP, coupled with very high resolution microscopy, is now making it possible to track the movements of individual transcription factors relative to the DNA elements to which they bind.[320] Such approaches will be vital in allowing scientists to study how changes in the cellular environment affect the expression of particular genes. The combination of these types of analysis has led to some major new insights. One is that even when chromosomes are in their uncondensed state, they generally occupy a defined region within the nucleus. But it has also become clear that, within such a chromosome 'territory', there are further levels of organization that divide the chromosome up into specifically defined structural and functional regions. In particular, recent studies have revealed an important sub-level of structure and function within the chromosome, so-called 'topologically associating domains' or TADs for short.[313] These are regions of the chromosome that can vary in size from a few hundred kilobases to several megabases. Within these regions there is a high level of local contacts, while their boundaries act as a barrier to contacts with other regions. This finding goes some way to addressing a conundrum that has puzzled biologists ever since enhancers were first discovered, namely how gene regulatory elements that act over such long distances do not activate any random gene in the genome. In fact, it seems that TAD boundaries act as 'insulators' that prevent the influence of an enhancer spreading beyond the TAD in which it is contained.[321] Such insulating effects are mediated by proteins like cohesin, previously shown to be essential for chromosome segregation in dividing cells.[322] This shows that although there has been a tendency in the past to separate the replication of the genome from its expression, some key factors clearly regulate both processes.

One issue still to be fully resolved in understanding how the genome functions as a 3D entity is determining what drives enhancer looping. Initially, it was assumed that simple random movement brought enhancers into contact with the genes they control. Yet this makes it hard to explain a surprising finding of ENCODE, which is that only in a small minority of cases—less than 10 per cent—do enhancers interact with the nearest gene in their vicinity as assessed in a purely linear fashion along the chromosome.[313] Instead, the growing consensus is that enhancer looping is an active process. One possibility is that 'bridging' proteins fill the gap between an enhancer and the gene it controls, and, indeed, a protein called 'mediator' seems to play such a role.[323] However, the recent surprising finding that enhancers are transcribed into RNA, and abolishing production of such RNAs inhibits expression of the genes they control, suggests that such RNAs may also be involved.[324] A key question to address now is whether enhancer RNAs are involved in forming DNA loops, possibly by guiding the enhancer to its target gene. Recent studies suggest this may be the case, but enhancer RNAs may also act in other, as yet unidentified ways, upon gene expression.[325]

The 3D structure of the genome also seems to play an important role in the process of splicing. Although this might seem counterintuitive, given that splicing occurs at the RNA level, it is becoming increasingly obvious that transcription of genes and splicing of the resulting RNAs are tightly coordinated, so that chromatin structure can have a major impact on splicing. Recently, John Mattick and his colleague, Tim Mercer, at the Garvan Institute, have shown, in a collaborative project with John Stamatoyannopoulos, that the portions of a gene coding for exons are far more accessible in the 3D genome than those that code for introns.[326] Moreover, exons destined to be selected for alternative splicing in particular cell types are also more accessible at the DNA level. Mercer believes this shows the genome is like 'a long and immensely convoluted grape vine, its twisted branches presenting some grapes to be plucked easily, while concealing others beyond reach. At the same time, imagine a lazy fruit picker only picking the grapes within easy reach. The same principle applies in the genome. Specific genes and even specific exons, are placed within easy reach by folding.'[327]

Although chromosome territories and TADs act to constrain gene expression in certain defined regions, there is also evidence that interactions between different regions of the genome can also sometimes occur over much greater distances, even between genes on different chromosomes. In particular, active genes from different chromosomes seem to congregate at sites called 'transcription factories' where substantial numbers of RNA polymerases and other enzymes involved in transcription are clustered. The idea of transcription factories was first proposed by Peter Cook and colleagues at Oxford University in the early 1990s.[328] They were using a labelled form of the nucleotide containing uracil, or U, found in RNA but not DNA, to visualize synthesis of RNA in the nucleus of a cell. In line with the idea that each individual gene is transcribed by its own RNA polymerase molecule, Cook and colleagues fully expected to see a homogenous distribution of labelling throughout the nucleus. Instead, they saw around three to five hundred concentrated clusters of transcriptional activity, and subsequent investigation with antibodies that recognized RNA polymerase proteins confirmed that these clusters contained such proteins. These findings led to the proposal that, rather than genes remaining stationary while the transcriptional machinery assembles around them, the situation is the other way around, with 'factories' composed of hundreds of RNA polymerases and associated enzymes being fixed at certain points in the nucleus while the genes to be transcribed come to them.[328]

The idea of transcription factories has been controversial since its first proposal, partly due to the difficulties of isolating such factories biochemically, and because there seems to be much variation between these entities in different cell types. However, the ability to capture interactions between different genomic regions has given a new lease of life to this idea by showing that genes from completely different parts of the genome that come together in the 3D nuclear space are associated with 'hot-spots' of transcriptional activity.[328] Tim Mercer believes this shows we need 'a new way of looking at things, one where the genome is folded around transcription machinery, rather than the other way around. Those genes that come in contact with the transcription machinery get transcribed, while those parts which loop away are ignored.'[327] Formation of hot-spots of activity is likely to involve a combination of biochemical compatibility between different regulatory proteins, effects of insulators, the chromatin environment, and the 3D architecture of the nucleus, all acting together.

As well as changing our understanding of the basic mechanisms of gene expression such findings also have clinical implications. In particular, a recent study of the genes that generate haemoglobin has confirmed both the importance of contacts between genes on completely separate chromosomes and the existence of transcription factories, as well as their significance in diseases where globin expression is abnormal.[329] Haemoglobin is, of course, crucial to our existence, carrying, as it does, oxygen to our cells and carbon dioxide away from them. The protein also has a distinguished role in the history of molecular biology since Max Perutz solved its 3D structure, making this the first such protein structure to be so determined. For his efforts Perutz received a Nobel Prize in 1962, the same year as Watson, Crick, and Wilkins received theirs. However, it's not only in advancing our knowledge of protein structure that haemoglobin has played a central role, but also our understanding of how genes are regulated. We've seen how, although enhancers were first identified in adenovirus, one of the first to be identified in our own genome was in the beta-globin gene.[330] In fact, this gene is just one of a cluster that includes the epsilon-, gamma-, and delta-globin genes, which code for embryonic and foetal forms of haemoglobin. Remarkably, the order of expression of these genes during development mirrors the order in which they occur on the chromosome, while studies have shown that a complex regulatory element—the locus of control region or LCR—controls the expression of all the genes in the cluster.[330]

Studying the expression of the beta-globin gene is important clinically because a number of serious genetic diseases affect haemoglobin. One such disorder is sickle cell anaemia, where a mutation in a single DNA nucleotide changes an amino acid on the surface of the protein to one of a different character. This alters the 3D structure of the protein, so that instead of forming a soluble protein, aggregates are formed that both compromise the oxygen-carrying capacity of the protein and also cause the sickle-shaped red blood cells characteristic of this disorder.[331] Another group of genetic diseases affecting haemoglobin are the thalassaemias.[332] These come in multiple forms but all involve abnormalities in expression of either the alpha- or beta-globin genes, which code for the alpha- and beta-globin proteins that make up haemoglobin in adult humans. Symptoms range from mild anaemia to fatal lack of properly functioning haemoglobin. Because mutations that cause thalassaemias can affect not only the protein-coding portions of the globin genes but also the regulatory regions that control them, studies of such mutations have led to important general insights about mechanisms of gene expression.[333] One unsolved issue is the question of how alpha- and beta-globin proteins are produced in evenly matched amounts, despite the genes coding for them being on completely different chromosomes. This is an important issue not only from a scientific point of view, but also for treatment of thalassaemias, because imbalances in the relative proportions of the alpha- and beta-globin protein chains can have toxic effects upon the cell.[332]

Studies investigating the interaction between different parts of the genome using cross-linking methods like those discussed, have shown that not only do the alpha- and beta-globin gene regulatory regions come into close contact, but they seem to be associated with the same transcription factory.[329] Understanding how this joint expression is regulated spatially in the genome could further our understanding of thalassaemias and lead to new ways to treat such disorders. In fact, this is just one case where understanding the 3D structure of the genome may aid diagnosis and treatment of disease. For instance, a breakdown in this 3D structure can occur during tumour formation. Cancer can be caused by activation of 'proto-oncogenes' or loss of function of 'tumour suppressor' genes.[334] Both classes of genes play important roles in regulating normal cell growth and division—it's only when they become mutated that they cause disease. We've seen how cancer can arise when a DNA region on one chromosome breaks and fuses with DNA from a different chromosome, such fusions being known as 'translocations'.[335] A cancer of the blood called Burkitt's lymphoma is triggered by a translocation in which the proto-oncogene MYC on chromosome 8, which plays a central role in cell growth, is brought into contact with the regulatory region for the immunoglobulin gene coding for antibodies on chromosome 14. Since antibodies are normally strongly expressed in B-lymphocytes of the immune system, this causes such cells to rapidly proliferate in a malignant fashion. Another translocation, known as the Philadelphia chromosome because of the city in which it was first identified, brings together two proto-oncogenes, BCR on chromosome 22 and ABL on chromosome 9, to form a fused protein BCR-ABL. This powerful stimulant to cell growth in white blood cells leads to a type of leukaemia.

Something that has puzzled cancer researchers for years is why certain gene combinations crop up so frequently in translocations. Now recent studies have shown that regions of the genome that fuse during translocations normally make close contact in the nucleus, and in the case of the MYC and antibody genes, are expressed by the same transcription factory.[336] So while breakage of chromosomes is a pathological process, its consequences will be partially determined by which genes are normally close to each other in the 3D nucleus. This has led Peter Cook and colleagues to suggest that characterization of the regulatory proteins involved in transcription factories, and how these change as a tumour develops, might lead to identification of new anti-cancer treatments.[337]

A breakdown of the 3D structure of the genome may also be one feature of the ageing process. So aged cells lack several key architectural proteins, and also have less condensed heterochromatin, than young cells. Further evidence of a link between ageing and genomic organization in the nucleus has come from studies of the premature human ageing disorder Hutchinson–Gilford Progeria syndrome.[338] Individuals with this disorder develop a wizened appearance and hair loss even as children, and generally die prematurely of a heart disorder or stroke. This disorder is caused by a mutation in the gene coding for lamin A, a protein involved in forming the nuclear lamina, and sufferers' cells show both a disorganized nuclear structure and also an absence of heterochromatin. How might a disorganized nucleus contribute to ageing? One possibility is that this exposes the genome to increased levels of DNA damage. Other premature ageing disorders in humans are due to defects in genes coding for 'DNA repair' enzymes that correct errors caused by UV radiation, environmental toxins, or simply mistakes made by DNA polymerase as it replicates the genome at each cell division. DNA that is less tightly organized in the nucleus may be more vulnerable to environmental insult, explaining why premature ageing is a feature of Hutchinson–Gilford Progeria syndrome.

Perhaps the most intriguing aspect of recent studies is the link identified between non-coding RNAs and long-range 3D interactions of the genome.[325] So although many non-coding RNAs have been shown to act locally to control expression of genes that they are located close to in the genome, there is also increasing evidence that some may act upon multiple targets that are much more distant. For instance, HOTAIR, a long non-coding RNA transcribed close to the HOX genes, members of the homeotic gene family, which, as we saw in Chapter 4, control body patterning, binds to 800 locations in the genome across multiple chromosomes.[325] Recent evidence suggests that this pattern of binding may be intimately linked to the 3D structure, but also that long non-coding RNAs in particular may play a central role as a kind of 'scaffolding' that ties different regions of the genome together both structurally and also in terms of function. That such RNAs, by virtue of their sequence but also 3D shape, can bind DNA, RNA, and proteins, makes them ideal candidates for such a role. Importantly, the scaffolds that they form seem highly dynamic, and this may be a key factor in the regulation of gene activity in a global fashion across the genome.

Such findings demonstrate the importance of genome structure in both the normal and pathological state. Importantly, they show that the view of genes as isolated entities strung out like beads on a string along linear chromosomes is a poor misrepresentation of the complex reality of the 3D genome. Nevertheless, such findings can still be reconciled with the idea of the genome as an essentially stable repository of information which is passed down through the generations with both its primary structure, and the information it conveys, being constant through those generations. But are genomes really that stable? It is time to take a closer look.

8

THE JUMPING GENES

'If you know you are on the right track, if you have this inner knowledge, then nobody can turn you off...no matter what they say.'   *Barbara McClintock*

'Jumping genes are fundamental because they're agents of change. Everybody knows that organisms evolve. What makes them evolve is that their genes are dynamic and in motion.'
                                                                        Nina Fedoroff

Imagine a map on which the key features moved about. This could make route-finding very difficult. If it were a map of England, one minute you might be heading south towards London, the next minute the capital city could have shifted north, next to York. Just as we expect the cities on road maps to maintain a constant position, so the scientists who first began to map the position of genes upon each chromosome did so with the justifiable assumption that, once located on the genetic map, those genes would stay in their allocated positions. This assumption may go some way to explaining the response when the geneticist Barbara McClintock announced in 1951, some two years before the discovery of the molecular structure of DNA, that portions of the genome could move about in the space of a few generations.[339] Unfortunately, despite the fact that McClintock had already made a name for herself establishing some key principles in genetics, such was the novelty of her findings that it took over three decades for them to be accepted by the scientific community. Even today, there is an ongoing debate about the functional significance of mobile elements, with the initial view that they are primarily parasitical entities only recently being challenged by evidence that they can play vital roles in gene regulation.

As a woman, McClintock faced many challenges in pursuing a career as a scientist in the early years of the twentieth century, not least from her own mother, who, when McClintock's father was away at the front during World War I, initially tried to prevent her daughter from going to college because of a fear that this would make her 'umarriageable'.[340] Luckily, McClintock's father interceded on his return from the war just before college enrolment began, and Barbara began the academic career that would result in several major discoveries in genetics and finally a Nobel Prize. McClintock never did marry or meet anyone she wanted to share her life with, but her work has transformed our view of the genome in ways that resonate to this day.

McClintock's first important scientific finding was the demonstration that the crossing over that takes place in meiosis during formation of eggs and sperm involves a physical exchange between each chromosome pair. As we saw in Chapter 1, Thomas Morgan surmised that this must be the case based on Frans Janssens' observations but he had never directly shown it to be true. But when, working with corn, McClintock identified a mutant that had a chromosome with an unusually shaped end, she realized that here was a way to directly test the idea. By crossing the mutant with a normal plant and then observing the inheritance pattern of characteristics associated with the odd-shaped part of the chromosome, in what has been called 'one of the truly great experiments of modern biology' McClintock showed that inheritance of such characteristics was always accompanied by a physical exchange of this part of the chromosome.[341]

McClintock continued studying genetics in corn for the rest of her career. Another major discovery she made using the plant was the identification of two key structural elements of chromosomes—'telomeres' and 'centromeres'—from the Greek for 'end' and 'centre', this being their respective locations on the chromosome. The discovery of telomeres was made when, inspired by Hermann Muller's success in using radiation to induce mutations, McClintock tried this approach in corn. One mutation with a very obvious physical form that she identified was so-called 'ring' chromosomes, which formed when the ends of a chromosome fused together. McClintock surmised from this that there must be a structure at the ends of chromosomes that normally stabilized them, and that this had been compromised by mutation.[342] These telomeres are like the plastic tips on the ends of shoelaces and prevent the chromosome from fraying just as such tips protect the lace. But telomeres also have a tendency to shorten every time a cell divides. This tendency is offset in young cells by an enzyme called telomerase, but as cells age this capacity diminishes, and this is one reason why cells in culture can only undergo a limited number of cell divisions before they die.[342] This limit can, however, be overcome in cancer cells which express telomerase at high levels. Telomerase is also active in eggs and sperm, so each new generation starts off life with re-lengthened telomeres.

In a recent study, mice were engineered to lack the telomerase enzyme and, as a consequence, suffered from advanced ageing.[343] However, when genetic engineering was used to reactivate telomerase in the mice for one month, surprisingly, not only did this stop the ageing, but it actually reversed the effects, so that the mice began to look significantly younger. As if they had drunk from Merlin's fountain of eternal youth, which, legend has it, still lies hidden in the Forest of Paimpont in Brittany, France,[344] shrivelled testes grew back to normal and the mice regained their fertility, the spleen, liver, and intestines recuperated from their degenerated state, and, in the brain, neural progenitor cells, which produce neurons and their supporting cells, were reactivated. Ronald DePinho of Harvard Medical School, who led the study, believes the findings show 'there's a point of return for age-associated disorders', while his colleague, David Sinclair, thinks that if a similar strategy could be used safely in humans, 'it could lead to breakthroughs in restoring organ function in the elderly and treating a variety of diseases of aging'.[345] However, David Harrison, who studies ageing at the Jackson Laboratory in Bar Harbor, Maine, believes 'telomere rejuvenation is potentially very dangerous unless you make sure that it does not stimulate cancer'. Harrison also questions whether mice lacking telomerase are a good model for human ageing, saying 'they are not studying normal ageing, but ageing in mice made grossly abnormal'.[345]

In contrast to telomeres, centromeres were identified by McClintock as bulges at the centre of each chromosome.[346] She noticed that these structures were always the first to line up at the centre of a cell before it began to divide, and it was the centromeres that seemed to guide each chromosome pair to an opposite pole of the cell. Such observations suggested that centromeres played a leading role in chromosome segregation. But the definitive evidence came yet again from the study of oddly shaped chromosomes that were the result of excessive irradiation with X-rays. After such treatment, some chromosomes had fused together so that they contained two centromeres. In this case, rather than being pulled towards one pole of the cell, the mutant chromosome was clearly being tugged in two opposing directions, providing the first direct evidence of the central role played by centromeres in the segregation process.

Normally, centromeres guide this process in a very precise fashion during the cell divisions of mitosis that occur during embryo development. If this were not the case, multicellular organisms like ourselves could never develop with the correct number of chromosomes in each of the several trillion cells that make up our bodies. However, mistakes do sometimes occur during the formation of eggs and sperm during meiosis. This can result in an organism with only one, or an extra copy, of a particular chromosome, in contrast to the normal situation in which each cell has two. In humans, these conditions are not generally compatible with life, but there are exceptions, particularly with the sex chromosomes.[347] So women with Turner's syndrome have only one X chromosome; such women have short stature, a broad chest, low hairline, and low-set ears, plus dysfunctional ovaries that normally result in sterility, as well as specific mental differences. In contrast, men with Klinefelter's syndrome inherit not one but two X chromosomes to go with their Y chromosome. These men have less muscular bodies, less facial and body hair, and broader hips; with such subtle differences, this condition often goes unrecognized. Genetically engineered mice with abnormal numbers of sex chromosomes are being studied to further understand the effect of such changes upon the body and help devise treatments for the associated symptoms.[347]

The most well-known disorder associated with more than two copies of a non-sex chromosome is Down's syndrome, caused by three copies of chromosome 21.[348] People with this condition have characteristic facial features and learning difficulties, as well as heart defects in later life and a shortened life span. One of the main risk factors for Down's syndrome is the age of the mother. This suggests that the machinery that drives each chromosome 21 to opposite poles of the cell becomes defective as women age. Current studies are focusing on the interaction between centromeres and this machinery, and how this changes with maternal age, in order to understand why the segregation becomes faulty.[348] Ultimately, such studies could result in treatments that prevent such events occurring.

Given such pioneering initial discoveries in McClintock's career, one might assume that the importance of her subsequent findings would be accepted as readily. Unfortunately, this was not to be the case, because the nature of her next discovery challenged the very basis of genetics as it was perceived at the time.

McClintock discovered mobile genomic elements while studying the genetics of colour in corn. As consumers, we generally buy corn that is uniformly yellow; however, many supermarkets now also sell fancy varieties in which the different kernels are multicoloured, and it was the genetics of these varieties that McClintock studied. However, as she studied the pattern of inheritance of colour across successive generations, McClintock came to a surprising conclusion—the chromosomal position of the genes associated with the different colours appeared not to be fixed, but instead seemed highly mobile.[339]

McClintock named this process 'transposition' and the mobile elements 'transposons'. She proposed that transposons were not the genes for colour themselves but rather their controlling elements, and she suggested that this could explain why complex multicellular organisms composed of cells with identical sets of genes can have cells with very different functions.[349] As we've seen, the distinction between structural genes and the regulatory elements that control them is generally associated with Jacob and Monod. Yet McClintock made this distinction in a paper published in 1953, seven years before the French scientists drew attention to it in their report on the lac operon in 1960.[350] So why was her pioneering suggestion essentially ignored at the time?

One problem was that McClintock's linking of this insight to the idea that such regulatory elements can move about the genome was just too implausible for most scientists at the time to accept. As she herself put it, after presenting her new proposal at a scientific conference, her findings were received with 'puzzlement, even hostility'. One characteristic response was that of Joshua Lederberg, who, after a visit to McClintock's lab, remarked 'by God, that woman is either crazy or a genius'.[351] There was perhaps some justification for this response. The picture of each chromosome as a linear map with genes aligned along it resonated with the common sense view of a map as a static entity upon which the main features— seas, rivers, mountains, valleys, towns, and cities—do not move. Of course, if we studied successive maps of the same area over time, we would see gradual change, both in natural features, and in the cities, roads, and other features constructed by our species. In the same way, our own genomes were recognized as subject to change, but in a painfully slow incremental fashion, as different genes within the genome were affected by mutation. Yet here was McClintock arguing that genomic elements could move about in a rapid fashion, in the space of a few generations. Such was the incomprehension and indeed hostility that McClintock encountered that, despite making other interesting new findings about the mobile elements—for instance that other 'suppressor' genes could inhibit their activity—she decided to stop publishing her work in this area. Instead, she diverted her studies into the origins of corn as a species, which she carried out during a series of trips to Central and South America where the plant originated.

But then, in the late 1960s and early 1970s, reports began to gradually filter in from other biologists of evidence for mobile elements in bacteria and yeast. Importantly, with new techniques for studying DNA at the molecular level discovered around this time, it finally became possible to show how transposition could occur. We now know there are two main types of transposons, both of which occur in the human genome (see Figure 20).[349] The first type replicate and then, just as text on a computer word file can be cut and pasted, they insert themselves elsewhere in the genome.[352] Although transposons of this type are no longer mobile in the human genome, they were active during the evolution of our

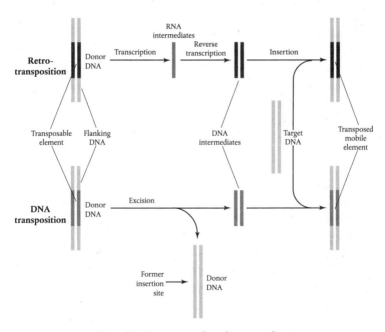

Figure 20. Transposons—how they move about

primate ancestors, about 37 million years ago.[353] The second type, which is still active in the human genome,[353] move via an RNA intermediate, which then turns back into DNA before it reinserts itself into the genome.[352] The recognition that RNA can code for DNA was one of the first challenges to Crick's central dogma that the information in DNA can only flow in one direction, via RNA to proteins. We saw in Chapter 3 how, since proteins are required to replicate and transcribe DNA, it could be equally valid to see information flowing back to DNA from proteins. Nevertheless, this is information flow in an indirect sense. In contrast, turning an RNA sequence into DNA is a very literal challenge to the central dogma, and this reversal of information requires a specific enzyme—reverse transcriptase.

In fact, reverse transcriptase was first discovered in quite a different context as a component of retroviruses, the most famous of which is HIV, the cause of AIDS.[354] Howard Temin of the University of Wisconsin, and David Baltimore of the Massachusetts Institute of Technology, made the discovery while studying RNA tumour viruses, this being another example, like the discovery of enhancers, where research into the link between viruses and cancer led to insights of great general relevance. An important clue as to how RNA tumour viruses work was the fact that, despite having an RNA genome, they have a permanent effect—tumour formation—on the tissues they infect. This led Temin to propose that since 'RNA is a transient molecule, so it must imprint itself on DNA'.[355] However, for years this notion was resisted because of the strength of the central dogma. Only when Temin and Baltimore independently isolated reverse transcriptase, the enzyme that carries out such imprinting by reversing the normal information flow, was the idea finally accepted. For the discovery Temin and Baltimore received a Nobel Prize in 1975. An important feature of RNA tumour viruses is that once conversion of the viral genome into DNA has taken place, it can now insert itself into the genome of the infected cell.[354] This ability to hide itself away in the genome of its host is one reason why HIV can be present in an infected individual for years before any overt symptoms are detected.

The discovery that retrotransposons also work via an RNA intermediate turned into DNA by reverse transcriptase came as a major surprise, for up till then this protein had been seen as something specific to retroviruses.[352] Such similarity has led to the suggestion that retrotransposons originally started life as retroviruses.[354] But it is also possible that retroviruses originated from retrotransposons. Deciding which version of events is correct is difficult without access to a time machine to replay evolution. However, recent studies indicate reverse transcriptase may have a very ancient origin, raising the possibility that this enzyme may date back to a time when the RNA world that we discussed in Chapter 5, was being replaced with one centred on DNA.[356] So, far from being a quirk of nature, reverse transcriptase may have been a key architect in the biological world that we now inhabit.

The discovery of the mechanism of transposition meant that McClintock's pioneering ideas were finally vindicated, and would result in her being awarded the Nobel Prize for Physiology or Medicine in 1983, the first woman to win that prize unshared, at the age of 81. But as welcome as this belated recognition of her achievements was, for McClintock it was still only a partial victory. For, in contrast to her proposal that transposons played central roles in gene regulation, an idea that now became dominant was that these were parasitical elements with no purpose other than to propagate themselves. Justification for such an interpretation came from surprising findings that were beginning to emerge from other studies of the genome. So, not only were transposons now identified in multicellular animals as well as in plants, they appeared to be a major component of the genome in both cases. The reason why transposition was first detected in corn became clear when it emerged that this plant has a staggering 85 per cent of its genome devoted to these transposons.[349] But many other organisms also have significant proportions of these elements in their genomes. Even before the sequencing of the human genome it was clear that our species had vast numbers of these elements and the genome project confirmed the proportion to be 45 per cent.[350,357]

Such abundance, as well as an apparently simple repeating structure, meant that the idea of these elements as useless parasites quickly spread. Indeed, it was this discovery that played a significant part in the notion that the genome was largely junk, with Orgel and Crick arguing in 1980 that 'the spread of selfish DNA sequences within the genome can be compared to the spread of a not-too-harmful parasite within its host'.[358] However, the idea of harmlessness was challenged by the suggestion that, since transposons could move about, this was potentially very dangerous junk. For what was to stop one of these elements inserting itself within, or close to, an important gene, thereby disabling the gene or activating it in a non-regulated fashion? Evidence that this can happen in humans has come from a study of the disease haemophilia, caused by a defect in the gene for clotting factor VIII.[349] The study showed that, in one patient, this gene had been disrupted by an insertion of a transposon. Other studies have shown that some cases of cancer have been triggered by a transposition event disrupting tumour suppressor genes or activating oncogenes.[359]

In general, though, our bodies seem to be very good at suppressing unwelcome transposon activity. So the suppressors that McClintock identified in corn have their equivalents in other species, including our own.[349] In particular, DNA methylation, which plays an important role generally in regulation of our genes as we saw in Chapter 6, is also important in keeping transposon activity in check. In addition, a recent study has shown that transposons themselves have evolved a mechanism for self-limiting their activity.[360] Transposition is driven by the enzyme transposase, and the study showed that once a certain number of copies of a transposon are created, transposase concentration rises to such a level that it begins to saturate its own binding sites. Ronald Chalmers at Nottingham University, who led the study, believes that this is in line with the idea that 'a successful parasite is not fatal to its host but lives in harmony with it'.[361] The effectiveness of such inhibitory mechanisms is such that one transposon isolated from fish genomes because of its potential usefulness for gene therapy was nicknamed 'Sleeping Beauty', because it had been 'reawakened' only by artificial means after a 'sleep' of 20 million years.[362]

One particular part of the body where transposon activity needs to be tightly regulated is in the genomes of sperm and eggs, for a transposon moving there to a new position could lead to sterility, the ultimate disaster from a Darwinian point of view. For this reason, the class of non-coding RNAs called piRNAs play a particularly important role in the gonads as suppressors of transposon activity. First discovered in fruitfly testicles by Alexei Aravin and colleagues at Moscow State University in 2001, they were subsequently shown to be present in the testicles and ovaries of all animal species, including humans.[363] piRNAs work by combining with piwi proteins, close relatives of the Argonaute proteins that are involved in RNA interference. The DNA sequences that code for piRNAs are bunched together in the genome in clusters. What was particularly surprising for researchers first investigating these clusters was the huge number of different piRNA sequences, there being at least 50,000 different varieties in mammals like ourselves, and maybe as many as 800,000, according to one recent study, in mice.[363] This diversity allows piRNAs to act like a mini immune system that tracks down and disables different transposons, just as antibodies do in the body as a whole.

New evidence suggests, though, that in some circumstances the mechanisms that protect our DNA from transposons are undermined. In particular, certain human brain disorders can be caused by inappropriate transposon activity. Amyotrophic lateral sclerosis (ALS), also known as motor neurone disease or, in the USA, Lou Gehrig disease, after the famous baseball player that it affected, is caused by a degeneration of the nerve cells of the brain.[364] This disease leads to increasing inability to move, speak, swallow, and breathe, and usually results in death within ten years from respiratory failure, although another famous sufferer, physicist Stephen Hawking, has lived with the condition for more than fifty years.

Studies have shown that this disorder, as well as other neurodegenerative conditions such as Alzheimer's, are associated with defects in a protein called TDP-43.[364] A recent study, led by Josh Dubnau at Cold Spring Harbor Laboratory, has shown that TDP-43 normally binds to transposons and keeps them inactive.[365] The study also suggested that neurodegeneration can be triggered by an 'awakening' of the dormant transposons to which TDP-43 normally binds. Dubnau believes this indicates that 'TDP-43 normally functions to silence or repress the expression of potentially harmful transposons. When TDP-43 function is compromised, these mobile elements become overexpressed.'[366]

Studies like these confirm the idea of transposons as potentially dangerous parasites whose activity must be kept in check so as not to disrupt normal cellular and bodily function. Recently, however, evidence has been accumulating that transposons may have more beneficial roles. One such role is in the evolution of mammalian pregnancy.[367] A study has identified an unusually high proportion of transposon-derived regulatory elements near to genes involved in the development of the placenta, the tissue that nurtures the developing embryo and foetus in mammalian species like our own, and which distinguishes us from egg-laying mammals like the duck-billed platypus.[367] The placenta is a unique structure that acts as a bridge between the mother and her developing offspring, with the connecting role facilitated by a protein called syncytin. When expressed in cell lines syncytin causes neighbouring cells to fuse, and in the placenta this property is central for the fusion of the cells of the mother and her child. Yet syncytin originally started life as a retroviral gene that became incorporated into the genome as a mobile element. Indeed, its mechanism of action reflects this origin, for in its original viral form the protein was located on the surface of the virus and helped the latter fuse with the cells it infected. These findings have led scientists studying this process to suggest that 'capture of a founding syncytin-like gene could have been instrumental in the dramatic transition from egg-laying to placental mammals'.[367]

A gene that began life as a transposon also plays a vital role in our immune system. As we've seen, a crucial way in which the body fights infection is through its ability to generate antibodies against a seemingly unlimited number of foreign molecules, or antigens. Although some of this diversity comes from alternative splicing, a far greater role is played by a rearrangement of the DNA in the immunoglobulin genes coding for such antibodies. Immunoglobulins are composed of four protein chains, two heavy chains and two light chains. Together they form a 'Y' shape, and the tips of the Y constitute the highly variable region that is subtly different in each different antibody.[368] A large part of this variability arises through a mixing and matching of a huge number of different sub-regions of the immunoglobulin genes. The protein mediating this rearrangement is an enzyme called RAG, which not only started life as a protein that regulates cutting and pasting of DNA transposons, but still plays this role today as it scrambles the different immunoglobulin gene sub-regions, in different permutations each time.[369]

Such examples might be viewed as important but still only isolated instances of the usefulness of transposition. However, a more controversial possibility now being debated is whether transposons have a far more central role in the process of gene regulation, just as McClintock originally proposed. This reappraisal of the role of transposons has emerged from global surveys of the regulatory elements in the genome conducted by projects like ENCODE. This has suggested that a surprising number of gene promoters originated as transposons.[349] So, at least 20 per cent of regulatory elements seem to have been derived from transposons, but the real figure may be significantly higher since the original character of these elements can often be hidden by mutation. Transposons have an obvious advantage in their potential to be co-opted into regulatory roles, since they already come equipped with sequences involved in gene expression. This is because these elements power their own movement in the genome and express genes that allow them to do this. Moreover, as transposons can also contain DNA sequences that respond to various hormones, this offers an intriguing possibility: that transposition might be strongly influenced by environmental pressures which lead to changes in the hormonal state of the organism. Such a possibility would be in line with McClintock's suggestion, first made in 1950, that environmental changes that stressed the organism, such as heat shock and starvation, can trigger an increase in transposon activity.

Indeed, an increasing number of recent studies in plants have confirmed this ability of stress to activate transposons. Moreover, there is increasing evidence that stress can lead to such transposition-led changes in the genomes of mammalian species.[370] For instance, one study showed that changes in transposon activity in the mammalian brain can be triggered by stress. Post-traumatic stress disorder is a condition whereby people exposed to traumatic events can become stressed or frightened even when no longer in danger. In order to gain insights into this condition, studies have been conducted on rats subjected to repeated stress. The rats subsequently show a disproportionate reaction, such as a tendency to freeze, even when placed in a non-stressful situation. When gene expression in such rats was analysed, greatly increased transposon activity was detected in a brain region called the amygdala that is known to play an important role in emotional responses and decision-making.[370]

Clearly, this is another example in which activation of mobile elements has a negative impact upon the body. But there is also evidence that transposon activity might have a more creative role. In particular, recent studies have suggested that the genomes of different cells in human individuals may be far more different than previously suspected.[371] It has long been an assumption in biology that the different cells of human beings, like other multicellular organisms, have the same genome, bar exceptions like red blood cells, eggs and sperm, and antibody-producing cells, which respectively have no DNA, only half the amount of this substance, and a scrambling of the immunoglobulin genes. However, there is increasing evidence, based on the ability to sequence the whole genomes of individual cells, that the DNA in different cells in the body may be far more divergent than thought.[371]

To some extent, such diversity is due to mistakes made by the replication machinery during the DNA copying process that occurs every time a cell divides, or by mutations that can occur when a cell is exposed to radiation or toxic chemicals from the environment.[371] This is a significant cause of cancer, as oncogenes are activated or tumour suppressor genes disabled, whether by too much exposure to the sun, triggering melanomas, or smoking giving rise to lung cancer. However, some of this diversity appears to be due to transposon activity and, intriguingly, this may play an important role in normal physiology. So, recent studies of the human brain have shown that active transposition is much more widespread than thought, suggesting that this might contribute to the plasticity which is such a characteristic feature of this organ.[370] And just as the immune system creates many genetically different antibody-producing cells to respond to foreign antigens, genome diversity in the nerve cells of the brain may allow it to respond to all the challenges that life throws at us.[371] It has even been proposed that such genetic diversity in the brain could be a significant cause of human individuality and explain why even identical twins can have quite different personalities and ways of dealing with situations.[370]

Such studies demonstrate that transposon activity may be important in the individual organism. However, what about another suggestion of McClintock's: that transposons play a central role in evolution? It is this that has proven to be the most controversial aspect of her work. McClintock argued that an increase in transposition might have benefits for the host organism facing environmental stresses, by initiating a rise in mutations that could provide an increase in variants for natural selection to act upon. Initially, this was thought unlikely since it was assumed that it could only occur in a fairly crude fashion, by disruption of gene function or non-specific activation of genes. However, McClintock always maintained there was a more creative side to transposition, and discoveries about the importance of transposons in forming new gene regulatory elements suggest she may have been right.

In addition, increasing evidence suggests that elevated transposon activity may accompany the origin of new species. For instance, a recent study has found that the timing of bat species' expansion coincided with an increase in transposon activity around 30 million years ago.[372] Moreover, this expansion also seems to have been linked to the creation of new miRNAs, which could have contributed to gene expression and therefore evolutionary novelty. David Ray of Mississippi State University, who led the study, believes this shows that 'transposable elements have the potential to shift evolution into overdrive by rapidly introducing large numbers of small RNAs. Those small RNAs don't change the proteins that genes code for but instead impact how and when the genes are expressed, thereby allowing for rapid changes in the way organisms interact with their environment.'[373] Other studies have suggested that increased transposon activity accompanied the origins of other vertebrate groups.[374] Perhaps most provocatively, Roy Britten has proposed that the high level of recent transposition events in human evolution, and the dramatic changes in that evolution, such as the rapid growth of the brain, are intimately connected.[375] This is a subject to which we will return later when we consider the role of the genome in the evolution of human consciousness.

As we've seen, one of McClintock's central beliefs was that transposition drove evolution by sensing changes in the environment, for instance, those that led to stress. But this led to the criticism that there was no obvious mechanism whereby environmental changes could be communicated to the genome. This was something McClintock herself never explained adequately. Aware of this, in her Nobel Prize acceptance speech in 1983 she challenged biologists 'to determine the extent of knowledge the cell has of itself, and how it utilizes this knowledge in a "thoughtful" manner when challenged'.[376] On another occasion she spoke about 'smart cells', meaning that the genome as a whole must have some way of sensing, evaluating, and responding to changes in the environment. Unfortunately, this type of language seemed, to many critics, to verge on mysticism, and prompted one to ask 'does the organism...have foresight, conjuring up just the kind of restructuring that the occasion demands?'[376]

The problem was that, in the mid-1980s, there was still no conception of how signals from the environment might affect the genome's activity in such a 'thoughtful' manner, and, indeed, such an idea was anathema to many biologists, for whom it conjured up a long disregarded figure from the past, someone whose ideas had supposedly been discredited many years previously by Darwin. This scientist—Jean-Baptiste Lamarck—has, since that time, been spoken of generally as a figure of fun, but developments over recent decades have begun to challenge that view.

9

THE MARKS OF LAMARCK

'It is not the…character and form of the animal's bodily parts, that have given rise to its habits and particular structures. It is the habits and manner of life and the conditions in which its ancestors lived that have in the course of time fashioned its bodily form, organs and qualities.'   *Jean-Baptiste Lamarck*

'We can't any longer have the conventional understanding of genetics which everybody peddles because it is increasingly obvious that epigenetics— actually things which influence the genome's function—are much more important than we realised.'   *Robert Winston*

Paris at the end of the eighteenth century was not a safe place for aristocrats, or indeed anyone with a link to the old order. The French Revolution was in full swing and heads were rolling. As such, Jean-Baptiste Pierre Antoine de Monet, Knight of Lamarck, and keeper of the herbarium at the former Jardin de Roi, the king's garden, had significant reasons to be worried for his life. True, Lamarck was a respected scientist, being one of the foremost biologists of his era; indeed, it was he who introduced the words 'biology' and 'invertebrate' into our vocabulary.[377] However, Antoine Lavoisier, discoverer of oxygen, had also been a highly respected scientist, yet that had not prevented him being guillotined for being one of the king's tax collectors. Lamarck, though, was a disciple of Jean-Jacques Rousseau, now dead but still a leading intellectual influence on the revolution, and he could see exciting possibilities for the future despite the dangers. Sitting by the fire one night with his wife Marie, Lamarck recalled an occasion five years earlier when they had gathered at a spot not far from where the Eiffel Tower now stands, to watch inventor Jean-Pierre Blanchard embark upon one of the first manned hot-air balloon flights. Now turning to Marie, Lamarck said, 'If I can endure, I will ignite something new. If one idea can ascend like the balloon then I will be remembered. That is enough.'[378]

Lamarck did have an idea that would make him remembered, but, unfortunately, for most of the next 200 years, it would be as an object of scorn and ridicule. This great idea was his proposal, 50 years before Charles Darwin, that life arose through a process of biological evolution. In fact, he was not the only person coming to such a conclusion; most notably, so was Darwin's own grandfather, Erasmus Darwin. Nevertheless, it was Lamarck who became most associated with this idea, so that Charles Darwin himself would later acknowledge that 'Lamarck was the first man whose conclusions on the subject excited much attention . . . he first did the eminent service of arousing attention to the probability of all changes in the organic, as well as in the inorganic world, being the result of law, and not of miraculous interposition.'[379] At a time when the established view was that God created the world and all its life forms in seven days, with human beings at the pinnacle, Lamarck's bold idea was as revolutionary as the period in which it originated. For it suggested that material forces alone, and not some supernatural creator, could explain the origins of humankind, as well as showing that we were not as different from other species as we thought. Perhaps this was why Lamarck's ideas drew such venom in the years following the revolution, as Napoléon came to power and steered French society in a less radical direction. In particular, the naturalist Georges Cuvier mounted a bitter opposition to Lamarck's evolutionary ideas, both in the latter's lifetime and also after his death. Indeed, in one of the most backhanded 'eulogies' ever delivered at a funeral, Cuvier used this opportunity to mock and criticize Lamarck's theory of evolution even as his opponent lay fresh in his coffin, accusing him of being someone who 'constructed vast edifices on imaginary foundations'.[380]

Cuvier was, however, helped by some obvious potential flaws in Lamarck's proposed driving forces of evolution.[381] One was a tendency for organisms to become more complex, moving 'up' a ladder of progress. However, it was Lamarck's suggestion that the environment directly influences the hereditary material that was singled out for attack. An example often used to illustrate this suggestion is a giraffe stretching its neck to reach the highest leaves on a tree, and thereby somehow giving rise to descendants with slightly longer necks, as if sheer force of will were capable of modifying the biology of subsequent generations. But, as critics pointed out, if evolution really did work in this way, then why are blacksmiths' children not born with muscular forearms or young giraffes with much longer necks than their parents?

Lamarck's response was that such changes would be too slow to reveal themselves in a single generation, instead taking 'many thousands of years'.[382] Unfortunately, such a timescale was hard to grasp for eighteenth-century citizens, who would have surmised from reading the Bible that the Earth itself was only 8,000 years old. And it would be over a century before it was conclusively shown that our planet is far more ancient, at four and a half billion years old, leaving plenty of time for evolutionary changes to take place.[383] But there was another problem with Lamarckism, namely the lack of an obvious material basis for Lamarck's two proposed evolutionary drives. Not only was it unclear why organisms should tend towards greater complexity but, in the absence of any understanding of the material nature of the hereditary substance, it was not easy to imagine how this could be affected by the environment. These problems, together with the challenge that Lamarckism represented towards religious orthodoxy, was one of the reasons why it was only popular in the early nineteenth century with radical groups, such as the Chartists in Britain.[384] This association with radicalism may be one reason why Darwin held back for so many years from publishing his own theory of evolution by natural selection, and why the socialist Wallace was less inhibited in putting forward his version of that theory.[385]

The theory of natural selection, in contrast to Lamarckism, provided a clear mechanism for evolution in proposing that the different capacity of variants in a population to survive explained why some species have evolved and others have become extinct. Yet, as we've seen, this theory suffered from its own mechanistic flaw as long as inheritance was assumed to involve a blending of factors originating from each parents' blood, which would cancel out the very variation that was supposed to drive natural selection. And although the way out of this impasse had already been identified by Mendel in Darwin's own lifetime, unaware of this, the latter became increasingly unsure of the primacy of natural selection, ironically appealing to Lamarckian mechanisms as additional factors in the last years of his life.

However, when the importance of Darwin's ideas was resurrected by the rediscovery of Mendel's work, and then by the 'new synthesis' of Mendelism and Darwinism, Lamarckism was once again marginalized. With the discovery that DNA was the hereditary material that acted like a linear code, the new orthodoxy was that the environment could only influence inheritance through the random generation of mutations in the DNA sequence, the variants produced by such mutations then becoming the raw material for natural selection to act upon. However, this is necessarily an indirect effect, in contrast to Lamarck's model, in which the act of a giraffe stretching for the highest leaves directly leads to an increase in neck lengths of subsequent generations. It also requires a far longer timescale than that envisaged by Lamarck. Recently, however, evidence has been emerging that Lamarck's proposal for a direct effect of the environment on the hereditary material may have not been so far off the mark at all, with one leading biologist, Eugene Koonin of the US National Institutes of Health, stating that 'Lamarck is back and perhaps with a vengeance'.[386] In particular, there has been a new recognition that gene activity may be altered in many ways that do not involve changes in the DNA sequence. What remains controversial, though, about such 'epigenetic' changes is whether their effects are only important over the lifetime of an organism or a few generations after, or instead have a significant influence on a longer timescale.

Although epigenetics has particularly flourished over the last decade, its origins go back much further. Indeed, the term was first proposed by British scientist Conrad Waddington in 1942; he created it by combining two terms, 'genetics' and 'epigenesis', the latter meaning the processes and events that bring the mature organism into existence.[387] In the 1930s Waddington travelled to the USA to work in Morgan's fly lab, where he began studying mutants as a way of understanding the mechanisms underlying embryo development. One mutant identified by Waddington had part of an antenna transformed into a segment of leg. He interpreted this as showing development is a series of branching decisions, regulated by the genes. In fact, we now know such changes are due to mutations in the homeotic genes, which, as we've seen, specify different bodily regions from head to foot and, remarkably, are lined up in the same order along the chromosome as the characteristics they impart to the embryo. In the absence of such molecular information, Waddington nevertheless came up with some interesting proposals. One was that developmental decision-making normally follows defined channels, with this 'canalization' meaning that a certain amount of genetic variation can be tolerated without obvious effects upon development, until a threshold is reached, at which point development can flip over into an alternative channel.[387]

According to Waddington, the development of the organism was not simply an outcome of the additive effects of all the individual genes, but rather their interaction with each other in a dynamic system, meaning that changes in expression of any one gene must be considered in terms of their effect on that system. This resistance of the body to genetic variation may explain an interesting phenomenon noticed by scientists studying 'knockout' mice engineered to lack expression of a particular gene. As we mentioned in the Introduction, surprisingly many knockouts have far less effect on bodily form or function than might be expected.[388] One explanation for this phenomenon is that many genes occur as members of families. If one gene is knocked out, another family member may take its place by a process of 'remodelling'.[389] In addition, many important cellular processes are regulated by multiple signalling pathways acting in parallel, so loss of one of these might be compensated for by a greater activity of a parallel pathway. What remains unclear is why some knockouts have very pronounced effects while others have little. In this respect, Waddington's canalization may operate to varying degrees for different processes depending on the extent to which compensation can take place. Importantly, as we'll see in Chapter 11, this concept may help us understand genetic diseases in humans.

At a time when the molecular basis of the genome was still being worked out, Waddington's theories represented an important insight into the complexity of the relationship between genes and the bodily characteristics they influenced. But Waddington's research was heading in an even more surprising direction. Investigating whether changes in the embryo's environment could affect its development, he discovered that exposing fly embryos to high temperatures during their development led, in a few cases, to the disappearance of a vein in the fruitfly wing.[387] Clearly, this aspect of development was susceptible to perturbation by such treatment. What was surprising, though, was that when flies with the missing vein were bred with each other, and the temperature shock and selection were repeated for a few generations, it was possible to create a population in which all flies lacked the vein.

Waddington identified other examples of this phenomenon. For instance, he found that treating fly embryos with ether could induce the formation of flies with four wings rather than two, and combining ether treatment of embryos and selection across generations eventually gave rise to only four-winged flies.[387] Such examples raised a number of important questions. Clearly, here was a very artificial situation in which Waddington himself was selecting offspring with particular characteristics. Yet the fact that an environmentally induced change could become heritable over a few generations suggested that something was happening to the hereditary material on a far shorter timescale than could be explained by spontaneous mutation.

Such a rapid influence of the environment seemed more in line with Lamarckism than with the new synthesis of Darwinism and Mendelism. But just as Lamarck's theory of evolution had foundered due to a lack of an obvious mechanism, so, in the 1940s, the lack of a clear idea about the material basis of the gene made it very difficult to even conceive how the environment could be acting in such a manner. If anything, the identification of DNA as the 'molecule of life' made Waddington's findings even harder to explain.[387] For if DNA operates as a digital code, and the only way this code can be altered is by mutations gradually changing the DNA sequence, then the speed of change in the examples discussed makes little sense.

As such, while the molecular biology revolution of the 1960s and '70s gathered pace, Waddington's findings were relegated, for many years, to the status of unexplained curiosities.[387] However, other evidence was emerging that suggested a far more direct influence of the environment on the hereditary material than would have been suspected, according to the standard orthodoxy of genetics, and some of it was in humans. Undoubtedly the most famous case of epigenetics affecting human health is the famine that affected Holland in the winter of 1944–45 at the end of World War II.[390] Because of a Nazi blockade on supplies entering the country, Dutch citizens suffered from an extreme lack of food. Such was the extent of the famine that starving people resorted to eating tulip bulbs and sugar beet to stay alive.[391] As it was, around 22,000 people died. However, the Dutch famine has become well known in scientific circles because of the surprising effects it had on women who were pregnant at this time, and their offspring. Strikingly, these fell into two categories. So women who were starved at the end of their pregnancy gave birth to children of a smaller birth weight. This was not so surprising; however, when girls born at this time grew up and themselves had children, these were also of reduced birth weight, despite their mothers having grown up in an affluent post-war society.[390] A different pattern was seen with women starved at the beginning of their pregnancy. While their subsequent access to food meant that they had offspring of normal birth weight, these children had a tendency to obesity, as if lack of food early in life had led to an urge to overindulge later in life by way of compensation. And in this case too, this tendency was passed on to their children.

Another historical example showing that food availability for one generation can affect not just children, but even grandchildren, of affected individuals, comes from Norrbotten, the northernmost county of Sweden, where the vagaries of the weather meant that, in the past, inhabitants could be subjected to famine, but also to periods of surplus.[392] Researchers studying whether such differences affected future generations found that grandchildren of men who had suffered famine lived longer than normal, while those of men who had lived through a time of surplus had a shorter lifespan. Such differences were linked to problems in cardiovascular health. So, in this case, a surplus of food was associated with detrimental effects on later generations. And since the effects were transmitted through men, this suggested that such effects were due to changes in the genome, not the womb.[392]

Such studies raised the question of how the environment might transmit such effects through the genome, given that they seem too rapid to involve a change in DNA sequence. In the end, a potential answer came from two discoveries that we discussed in Chapter 6, namely that DNA can be modified by methyl groups, and that the histones that wrap around DNA can also be modified chemically, for instance, by the addition of an acetyl group.[393] While acetylation of histones directly affects the tightness with which histones bind the DNA, thereby making the DNA more accessible to gene regulatory proteins, other histone modifications serve as a recognition signal for proteins, which then act to influence gene expression.[394] The proteins involved in epigenetic signalling have become known as 'writers' and 'readers': the former deposit epigenetic marks, while the latter interpret those marks and carry out the associated regulatory function. Other proteins act as 'erasers', by removing epigenetic marks. This reversibility of epigenetic changes is one crucial way in which they differ from genetic ones, since once a mutation occurs in DNA, this becomes a permanent feature. Why such proteins target one region of the genome over another has been unclear; however, recent studies suggest non-coding RNAs are central to this process.[395] This is part of an emerging dual aspect of such RNAs which can both recognize DNA elements due to sequence similarity, as well as co-opt proteins that modify DNA and its accompanying histones, thus directing epigenetic enzymes to specific gene targets.[395]

That epigenetic changes play key roles in our cells, there is now no doubt. We've already discussed the conundrum whereby a single cell, the fertilized egg, can develop into an organism of thousands of different cell types, each with a different function, yet, in general, such different cell types contain the same genome. The discovery of transcription factors went some way to explaining this conundrum because it showed that different types of such factors in a cell would affect which genes were turned on or off. However, increasing evidence also implicates an important role for epigenetic changes.[396] Initially, epigenetic changes were thought to define cell specificity in a purely negative fashion, by preventing access of transcription factors to a gene; however, epigenetic features such as histone modifications are increasingly viewed as acting in a more positive, dynamic fashion.[397] In addition, epigenetic changes may affect the position of genes and their regulatory regions within the 3D nucleus, which, as we've seen, can greatly affect whether genes are turned on or off.

One curious feature of epigenetic changes during embryo development specific to mammals is a phenomenon called imprinting.[398] This involves certain genes being switched on or off, depending on whether they come from an individual's mother or father. The consequences of imprinting have been known since ancient times, when it was realized that animal hybrids generate different types of offspring depending on which species is the mother, and which is the father. So mating a male horse and a female donkey produces a hinny, while the opposite combination generates the more common mule. This suggested that there must be something different about the male and female genomes, for otherwise it is difficult to see why the two combinations should generate different types of animal.

The first experimental demonstration of this difference was made by Azim Surani at Cambridge University. In the 1980s Surani decided to investigate whether the genomes of two sperm, or two eggs, could develop normally when transplanted into an egg that had its own genome removed. If the two genomes were essentially the same, this ought to have resulted in the development of a normal embryo. Instead, Surani found that both combinations led to highly abnormal development.[398] He proposed that the male and female genomes must be modified, such that balanced development was only possible if both were present. Confirmation that this was the case came with the discovery that certain genes had a different pattern of methylation depending on whether they came from the father's or the mother's genome, which also affected whether they were turned on or off.

Why does such a phenomenon occur in mammals? One theory builds upon the fact that mammalian females bear their young internally, as opposed to laying eggs.[398] Because male mammals only have to invest a sperm in producing a new embryo, there is an evolutionary incentive for the paternal genome to boost growth of the embryo. However, female mammals must nurture the developing embryo inside their bodies at considerable cost to themselves; indeed, this could even become life-threatening should the new life form demand too many resources. The theory therefore predicts that genes switched on in the father's genome should boost embryo growth, while those in the mother suppress it. And, indeed, in general this seems to be the case. However, an opposing theory, for which there is also some evidence, has proposed that imprinted genes act cooperatively to optimize foetal development and the well-being of the mother. Given that at least 150 imprinted genes have been identified, with quite different characters, it is probable that elements of both theories may be correct.

Imprinting has been linked to a number of human diseases. In particular, genetic defects in imprinted genes can cause completely different symptoms depending on whether the disorder is inherited through the mother or father.[399] So Prader–Willi syndrome is associated with various symptoms, but the most prominent is an insatiable appetite, leading to life-threatening obesity. In contrast, individuals with Angelman syndrome have severe learning disabilities, jerky movements like hand-flapping, and engage in frequent laughter and smiling. Yet, despite their completely different characters, both syndromes are due to a loss of the same region of chromosome 15. However, while Angelman syndrome is caused by loss of expression of an imprinted gene that is normally only on in the maternal genome, Prader–Willi symptoms are due to an absence of expression of a neighbouring gene that is generally only on in the paternal genome. As well as playing a role in these severe disorders, there is increasing evidence that subtle differences in expression of imprinted genes can contribute to more common disorders like obesity, diabetes, psychiatric illness, and cancer.[398]

The role of epigenetic changes in embryo development explains why cloning remains a very inefficient process. For not only must the newly introduced genome come into contact with a whole new set of transcription factors, for these to influence its expression it must also undergo a fundamental change in its chromatin state. That such 'remodelling' can happen on a vast scale is remarkable in itself, but the fact that cloning is only successful in a small minority of attempts shows that it is a far from assured event.[400] The examples just given also show the importance of epigenetic changes for normal development. But what relevance do such changes have to Lamarck's proposal that the environment can shape the hereditary material? In fact, a growing number of studies have shown that the epigenetic state of the genome is more responsive to environmental influences than previously suspected. One important influence on the epigenome is diet.[401] Substances ranging from green tea, garlic, carrots, broccoli, and cumin, can all affect the methylation state of different genes. The influence of diet on the epigenome is an active area of study, since it might point to ways to improve our health through the manipulation of diet, but also show whether consumption of cheap food full of fat and sugar—so-called 'junk food'—affects more than our waistlines.

Epigenetic changes also seem to be an important part of the body's response to stress. This response is mediated by a rise in 'stress hormones' like cortisol. Such hormones are released from the adrenal glands, these being stimulated by a region of the brain called the hypothalamus acting via the pituitary, the three together forming the HPA axis.[402] Stress hormones are members of the 'steroid hormone' family because of their chemical structure. Steroid hormones mediate their effects in the body by switching on certain target genes: to do this, the steroid hormone enters the cell across the cell membrane, and binds to a receptor inside the cell's cytoplasm. The combined hormone and receptor then effectively becomes a transcription factor that enters the nucleus and activates its target genes.

The effects of stress hormones on gene expression will persist as long as levels of these hormones in the blood remain high, but it was always assumed that, once they fell, so would the changes in gene expression. However, recent studies suggest that stress can cause more long-lasting epigenetic changes in the genome.[402] So, baby rats raised by mothers with a defect in their ability to look after their young grew up to have higher levels of methylation of the regulatory region, and so lower activity of the gene coding for the cortisol receptor. This effect was environmental, being also seen with rats born to biological mothers who did care for their young, but which were then fostered by uncaring mothers. A higher level of DNA methylation was also seen in the cortisol receptor gene in human suicide victims who had been abused as children.[402] In both rats and humans, therefore, stress early in life seems to desensitize the response to stress hormones, and this is mediated by epigenetic changes.

Recently, the effects of stress on the human genome have been shown to be more dramatic than had been imagined. We saw in Chapter 8 how the ends of chromosomes are protected by DNA sequences called telomeres that shorten each time a cell divides, and whose gradual loss in a person's lifetime are one cause of ageing. However, a recent study has shown that telomeres can shorten much more rapidly in children exposed to extremely stressful situations. The study examined two sets of 9-year-old boys, one being children who had grown up in a poor and unstable environment, the other being boys from more privileged backgrounds.[403] The first set typically lived with a single mother who had multiple partners, and had been exposed to domestic violence and other types of stress. An examination of telomere length in cells isolated from the two groups revealed that some of the boys in the stressful home environment had telomeres that were a staggering 40 per cent shorter than normal. However, this was only true of some of the boys, and further analysis revealed that affected individuals had differences in the genes coding for dopamine and serotonin, two brain neurotransmitters. Although these chemicals play vital roles in mediating human characteristics like love, happiness, self-confidence, and motivation, imbalances in the levels of these two substances in the brain are associated with depression, bipolar disorder, and schizophrenia. Given that shortening of telomeres has been linked to ageing and susceptibility to disease, these are worrying findings. But they also suggest that the link between stress and the epigenome is a complicated one, and may explain why, although stressful environments may trigger mental disorders, biological differences may also decide which individuals are most at risk and therefore in need of rapid intervention.

Findings such as these suggest that our genomes can be influenced by the environment in much more direct and dramatic ways than suspected, which is a vital necessary element if there is to be any truth in Lamarck's version of evolutionary change. However, a more controversial issue that remains to be properly addressed is the question of whether such epigenetic changes can be passed down to future generations, as Lamarckism requires, and to what extent this shapes evolution in the long term. In this respect, examples such as the Dutch famine offer tantalizing suggestions that the environment can influence future generations, but what is the evidence that this is linked to epigenetic changes? Studying such questions in humans is necessarily difficult, both in terms of obtaining tissue samples to analyse and of tracking individuals across generations, which pose practical and ethical problems. However, studies of laboratory animals are helping our understanding of this issue. Such studies have shown that pregnant mice exposed to different diets or environmental toxins can pass on epigenetic changes, not just to sons and daughters, but also to grandchildren and great-grandchildren.[399]

Undoubtedly the biggest challenge for the idea that epigenetic changes can be transmitted across generations in mammals, including humans, is the existence of mechanisms that erase previous epigenetic marks in the egg and sperm and impose new ones. Because of this, from a genomic point of view at least, each new generation was, until recently, thought to start out as a blank slate. However, the notion that this process is an absolute one is now being challenged. So, previously it was thought that all the histones in the sperm genome were replaced by protamines; however, as we saw in Chapter 7, it now appears that as much as 15 per cent of human sperm DNA is associated with histones. Moreover, a recent study has shown that paternal diet affects the chemical modifications of such histones, which may therefore such carry such epigenetic marks into the next generation.[404] There is now also evidence that protamines themselves not only act to protect the sperm DNA during its journey to the egg, but may also carry epigenetic information into the embryo. In addition, there is increasing evidence that far more genomic regions than thought may escape the erasure of DNA methylation that occurs in epigenetic 'reprogramming' during sperm development.[405]

Perhaps most remarkable are recent studies indicating that some non-coding RNAs can be transmitted to the next generation via the sperm, and that these may guide the placement of epigenetic marks. One such study found that male mice subjected to stress as babies, produced offspring that showed depressive behaviour and a tendency to underestimate risk. Analysis of the sperm of the stressed mice showed that they contained an abnormally high expression of five miRNAs,[406] one of which, miR-375, had previously been linked to the stress response.[402] Remarkably, not only were the immediate offspring affected, but also the grandchildren of the stressed mice. Both offspring and grandchildren had abnormal levels of the five miRNAs in their blood and in the hippocampus, the latter being involved in both memory formation and the mediation of stress responses. To discount the possibility that the effects of stress were transmitted socially, the researchers isolated RNA from the sperm of the stressed mice and injected this into fertilized eggs from unstressed parents—this also led to offspring with depressive behaviour and abnormal metabolism, characteristics that were passed on to subsequent generations.[406] This suggests that the effects of stress on future generations can be directly transmitted by sperm miRNAs.

How could stress lead to changes in miRNA levels in the sperm? One possibility is that stress hormones circulating in the blood make their way to the testicles, and trigger expression of miRNAs via stimulation of surface receptors on the sperm. However, an even more direct potential route has been recently identified, since miRNAs contained within 'exosomes'—membrane bound particles—have been observed entering sperm in the epididymis, the structure in which sperm are stored after they leave the testicles, prior to ejaculation.[407] It is possible, therefore, that miRNAs produced elsewhere in the body, for instance, the brain, could subsequently end up in the sperm and in the fertilized egg and embryo, providing a direct connection between the brains of one generation and the characteristics of future ones.

A major unresolved question is what relevance the propagation of epigenetic marks has for longer-term evolution. One interesting possibility is that such marks may facilitate more permanent genetic change. Such a possibility is based upon the discovery that methylated nucleotides are more prone to mutation than normal ones. Thinking along such lines, Eugene Koonin has recently argued that evolution may follow a 'two-phase process', with the first phase being the Lamarckian epigenetics and the second phase Darwinian selection of mutations'.[408] His proposal is that this would be akin to 'probing the waters...with epigenetic adaptation followed by the long-term genetic inheritance of the same adaptation should the challenge prove to be long-lasting'. If true, Koonin believes that this 'defies the common belief that evolution has no forecast'.[409]

Another possibility is that epigenetic changes make the genome more liable to transposition. As we've seen, stress may enhance transposition and, intriguingly, this seems to be linked to changes in the chromatin state of the genome, which permits repressed transposons to become active. It would therefore be very interesting if such a mechanism constituted a way for the environment to make a lasting, genetic mark. This would be in line with recent suggestions that an important mechanism of evolution is via 'genome resetting'—the periodic reorganization of the genome by newly amplified mobile DNA elements, which establishes new genetic programmes in embryo development. New evidence suggests such a mechanism may be a key route whereby new species arise, and may have played an important role in the evolution of humans from apes.[410] This is very different from the traditional view of evolution as being driven by the gradual accumulation of mutations. Instead, it suggests that the genome is built 'Lego-like out of codons specifying protein domains', and evolutionary change is 'largely a matter of nonrandom codon reorganization by natural genetic engineering mechanisms like retrotransposition'.[411] Such a viewpoint would be in line with the Danish embryologist Søren Løvtrup's belief that 'evolution is not a question of making new materials, but rather of using old materials for new purposes'.[412] Such issues remain to be resolved. But our new understanding of the genome and epigenome are challenging our view of both disease and what it means to be human, and it is to these matters that we will shortly turn. However, first it is time to step back and address once more the question of how much functionality there is in the genome in light of the new information that we have gathered about the different levels of genomic activity. For, as well as being a continuing topic of controversy, this issue has a significant bearing on what will follow in the rest of this book.

10

CODE, NON-CODE, GARBAGE, AND JUNK

'Inspect every piece of pseudoscience and you will find a security blanket, a thumb to suck, a skirt to hold. What does the scientist have to offer in exchange? Uncertainty! Insecurity!'   *Isaac Asimov*

'I think people get it upside down when they say the unambiguous is the reality and the ambiguous merely uncertainty about what is really unambiguous. Let's turn it around the other way: the ambiguous is the reality and unambiguous is merely a special case of it, where we finally manage to pin down some very special aspect.'   *David Bohm*

In one of his more philosophical moments, which, being a Frenchman in the existential 1960s, was quite often, Jacques Monod said that 'in science, self-satisfaction is death...it is restlessness, anxiety, dissatisfaction, agony of mind that nourish science'.[413] This is an interesting reflection, given that a common conception about science is that it is primarily about assured facts. This viewpoint is not surprising when we consider how science is generally taught in school, and even to an extent at university, using textbooks or course notes in which scientific facts are displayed as things to be learned, not as objects to debate or dispute. Now, in a sense, there is a good reason why certain scientific claims are taught as facts, namely that, having been around for a substantial period of time and having been subject to ample efforts to disprove them, it is assumed that they correspond to real truths about the world. It is because of this that Euclid's principles of geometry, Newtonian mechanics, and Darwin's theory of natural selection are all accepted as central cornerstones of modern scientific thought.

It would be a mistake, though, to assume that such certainties characterize science at its most cutting edge, since, by definition, any true foray into the unknown must be highly uncertain. In discussing the ENCODE findings, we saw that a key question that polarized opinion about the significance of the findings was the question of how much of the genome could be considered 'functional' and how much was 'junk'. Having now looked in detail at novel features of the genome emerging from recent studies, namely its 3D character, the mobility of genetic elements, and links between genes and the environment mediated by epigenetic mechanisms, it is time to reassess this issue of genomic functionality. Here, however, we face a problem, which is that the approach used to assess such functionality has a substantial influence on the outcome of this assessment.[414] And since the choice of approach is itself influenced by whether one subscribes to a traditional picture of the genome, or a more radical one, this is an issue likely to be characterized by uncertainty for some time in the future.

As we discussed in Chapter 6, a traditional way of assessing functionality in the genome is to assume that important functional elements are those that show sequence conservation between different species. The idea is that such regions of the genome have survived 'purifying selection', the tendency of natural selection to weed out portions of the genome that have a non-functional role. So, a typical approach would be to compare the human genome with those of different mammalian species and identify how many nucleotide bases are retained between these two genomes. Chris Ponting and colleagues, who recently made such a comparison, have likened it to training 'a time-lapse camera on a single nucleotide position in your genome and, by winding back time, watch[ing] how it changed by chance mutations as it was passed back through the generations (and along the germline) over hundreds of millions of years'.[415] If a nucleotide is functional, then it should change only rarely. This is because change is mainly detrimental to survival and therefore less likely to have been propagated to subsequent generations.[416] However, if the nucleotide is not functional 'changes would not have been selected against and thus would have occurred more frequently'.[417]

Such a comparison between mice and humans reveals that while there is extensive sequence conservation in the protein-coding genes, around 85 per cent, conservation outside these regions is far less. Within these non-coding regions of the genome, however, certain elements stand out as being conserved, in particular regions linked to genes that are binding sites for transcription factors and sequences that produce miRNAs, in line with the important roles identified for both these types of molecules as regulators of gene expression.[416,418] However, the remainder of the non-protein-coding genome showed far less conservation between mice and humans, such that only 3 per cent of the genome seems to be functional when assessed by these criteria.[418] This could imply lack of a functional role, although one problem with comparisons between the mouse and human genomes is that this might not detect genomic regions that are specific to our species, but might nevertheless have an important functional role. To address this issue, Ponting and colleagues also compared more closely related species, such as chimps and other primates, as well as looking at differences between different human individuals. When assessed by such criteria, this suggested that around 9 per cent of the human genome is functional.[418]

In contrast, another way of assessing functionality in the genome involves measuring its biochemical activity. This was the approach taken by the ENCODE researchers and, as we've seen, this has meant assuming that detection of transcription factor binding, DNA methylation and histone tags, and generation of non-coding RNAs, are all evidence of function. It was by this approach that ENCODE researchers came up with their high figure of 80 or more per cent functionality. Clearly though, this leaves us with a conundrum. So, while the estimate of 9 per cent functionality by Ponting and colleagues is a lot greater than the 2 per cent of the genome previously thought to be functional, it is clearly a lot less than 80 or more per cent. Is there any way of reconciling these two quite different figures? It is here that the debate becomes most heated. Those who believe the lower figure is the correct one argue that assuming biochemical activity equals function ignores the possibility that such activity might be just 'noise' and is impossible to reconcile with the apparent much lower levels of conservation. In contrast, those proposing a much higher figure believe that conservation is an imperfect measure of function for a number of reasons. One is that since many non-coding RNAs act as 3D structures, and because regulatory DNA elements are quite flexible in their sequence constraints, their easy detection by sequence conservation measurements will be much more difficult than for protein-coding regions. Using such criteria, John Mattick and colleagues have come up with much higher figures for the amount of functionality in the genome.[419] In addition, many epigenetic mechanisms that may be central for genome function will not be detectable through a DNA sequence comparison, since they are mediated by chemical modifications of the DNA and its associated proteins that do not involve changes in DNA sequence. Finally, if genomes operate as 3D entities, then this may not be easily detectable in terms of sequence conservation.

Another potential problem in the way that sequence conservation has been used as a measure of functionality in the genome is the fact that such conservation needs to be measured against a reference point, which, in this case, is the repetitive DNA sequences that have accumulated in the genome through transposition. These sequences are assumed to be useless, and therefore their rate of mutation is taken to represent a 'neutral' reference; however, as John Mattick and his colleague Marcel Dinger, of the Garvan Institute, have pointed out, a flaw in such reasoning is 'the questionable proposition that transposable elements, which provide the major source of evolutionary plasticity and novelty, are largely non-functional'.[420] In fact, as we saw in Chapter 8, there is increasing evidence that while transposons may start off as molecular parasites, they can also play a central role in the creation of new regulatory elements, non-coding RNAs, and other such important functional components of the genome.[421] It is this that has led John Stamatoyan-nopoulos to conclude that 'far from being an evolutionary dustbin, transposable elements appear to be active and lively members of the genomic regulatory community, deserving of the same level of scrutiny applied to other genic or regulatory features'.[422] In fact, the emerging role for transposition in creating new regulatory mechanisms in the genome challenges the very idea that we can divide the genome into 'useful' and 'junk' components. A point that Sydney Brenner once made in reference to junk DNA was to distinguish between 'the rubbish we keep, which is junk, and the rubbish we throw away, which is garbage ... everyone knows that you throw away garbage. But junk we keep in the attic until there may be some need for it.'[423] That previously useless items may take on new and important uses is important to bear in mind, against the idea that the functions of different genomic regions are fixed and unchanging.

The potential pitfalls of writing off elements in the genome as useless or parasitical has been demonstrated by a recent reconsideration of the role of pseudogenes. As we discussed in Chapter 4, these mutated, dysfunctional versions of protein-coding genes have traditionally been held up as a prime example of genomic garbage; indeed, it was in reference to pseudogenes that Ohno first coined the phrase 'junk' DNA.[424] Yet recent studies are forcing a reappraisal of the functional role of these genetic 'duds'.[425] One particular type of pseudogene is the 'pseudoenzyme'. We mentioned in Chapter 1 that enzymes are the class of proteins that catalyse chemical reactions in the cell, and perform jobs that include transport of foodstuffs into the cell, transformation of these into energy, and regulation of the genes coding for all these processes, as well as a diversity of other cellular functions. Central to how enzymes work is their 'active site': the specific region in the enzyme in which catalysis take place. These active sites are characterized by a very precise amino acid sequence, which forms a similarly precise 3D structure in which catalysis takes place. Indeed, this precision is the key to how different enzymes are able to specifically catalyse their own unique chemical reaction amidst the hundreds of thousands of other reactions, all taking place simultaneously in the cell and organism.

Given the importance of enzymes in the body, it was not surprising that one of the first tasks of geneticists, following the completion of the Human Genome Project in 2003, was to catalogue all the genes that code for enzymes in the genome. Yet what was surprising was how many of those identified seemed to be catalytically inactive, as defined by the presence of debilitating mutations in their active sites.[425] So, of the 518 human kinases, enzymes which, as we saw in Chapter 3, activate other proteins by adding a phosphate group to them, around 10 per cent lacked at least one of three key amino acids necessary for catalysing the phosphate transfer. In another class of enzymes that modify proteins by adding a sulfate group, more than half the members of this class seemed to be catalytically inactive.[425] All of this confirmed the idea that the genome was littered with the remnants of 'dead' proteins. And although knowledge about the existence of pseudogenes was nothing new, the surprising number of pseudoenzymes came as somewhat of a shock to genome researchers, so much so that one of them, Gerard Manning at Genentech, a biotechnology company in California, recalls that 'we thought we must have got it wrong'.[426] Recently, though, as in a zombie horror movie, the dead are showing surprising signs of life.

In particular, recent studies have shown that just because pseudoenzymes are catalytically inactive, this doesn't prevent them playing important roles in the cell. In general, this involves the regulation of a 'living' cellular partner.[425] Why pseudogenes often have such an active partner relates to how they come into being in the first place. So, a typical way in which pseudogenes form is through the duplication of a functional gene. Mutation of one member of the pair can disable it, while leaving its partner catalytically active. Initially, this was thought to be the end of the matter, but then biologists studying the evolution of pseudoenzymes became intrigued by the fact that the DNA sequences coding for some of them had changed little over millions of years of evolution. This suggested that these proteins had some function, for, as Patrick Eyers of Sheffield University, England, who was studying this question, puts it, 'biological systems don't bother keeping these proteins unless they are doing something important'.[426] In fact, subsequent studies have shown that pseudoenzymes have resisted change precisely because they have a variety of important roles assisting their catalytically active partner.[425] Such roles can include helping their partner to catalyse its specific reaction by forcing it into the correct shape, or acting as a bodyguard to transport it safely to its required location in the cell. The ability to play such specific roles is directly linked to the similarity between pseudoenzymes and their active partners since this allows the two to associate; yet, at the same time, the fact that pseudoenzymes are no longer required for catalysis has opened up a space for them to evolve in a variety of different ways that enhance their possibilities as regulatory agents.

The study of pseudoenzymes is not only driven by scientific curiosity: there is also now interest in this class of proteins as targets for new therapeutic drugs.[425] Pseudokinases in particular are being investigated as potential drug targets. This is because many kinases themselves are highly important targets for anti-cancer drugs, reflecting the role of this class of enzymes in normal cell growth but also in tumour formation. Such kinase inhibitors have been very successful, accounting for nearly $11 billion in sales in the USA alone. However, one drawback of such drugs is that because they target the enzyme's active site, and since this is fairly similar in different kinases, an inhibitor that targets one kinase can affect the activity of a different type, leading to unwelcome side-effects. So, although one of the most successful anti-kinase drugs, Gleevec®, has been effective in combating one form of leukaemia, it also causes abdominal pain, nausea, and fatigue. In contrast, because pseudokinases work through regions other than the active site, there is a hope that it might be possible to interfere with their activity, and therefore that of their catalytic partners, in a way that does not affect other types of kinases.[425]

Such new findings about the role of genomic elements previously presumed to be junk provide an important caution to the idea that we can simply write off such elements. However, while this suggests a qualitative shift may be required in our understanding of genome function, it does not directly address the quantitative issue of whether claims of 80 or more per cent of the genome are warranted. In this respect, one issue that has puzzled scientists for many years is the so-called 'C-value paradox'. This was the term coined by C. A. Thomas of Harvard Medical School in 1971, to refer to the emerging evidence at that time that studies of the amount of DNA in the cells of different species seemed to bear no relationship to their complexity as organisms.[427] So, as Ford Doolittle of Dalhousie University, Canada, has noted, 'humans have a thousand times as much DNA as simple bacteria, but lungfish have at least 30 times more than humans, as do many flowering plants'.[428] At the other extreme is the *Fugu* pufferfish—a species prized as a delicacy in Japanese restaurants but so toxic that, if prepared incorrectly, it can rapidly result in death.[429] But *Fugu* is also of great interest to biologists because its genome is unusually compact—clocking in at a mere 400 million bases, compared to our own 3 billion bases. Yet, despite having a genome only one eighth the size of ours, *Fugu* possess a similar number of genes. This disparity raises questions about the wisdom of assigning functionality to the vast majority of the human genome, since, by the same token, this could imply that lungfish are far more complex than us from a genomic perspective, while the smaller amount of non-protein-coding DNA in the *Fugu* genome suggests that loss of such DNA is perfectly compatible with life in a multicellular organism.[428]

Not everyone is convinced about the value of these examples though. John Mattick, for instance, believes that organisms with a much greater amount of DNA than humans can be dismissed as exceptions because they are 'polyploid', that is, their cells have far more than the normal two copies of each gene, or their genomes contain an unusually high proportion of inactive transposons.[430] Mattick is also not convinced that *Fugu* provides a good example of a complex organism with no non-coding DNA. Instead, he points out that 89 per cent of this pufferfish's DNA is still non-protein coding, so the often-made claim that it is an example of a multicellular organism without such DNA is misleading.[431]

Perhaps one of the strongest arguments against the relatively small degree of sequence conservation in the genome being an accurate reflection of the true extent of functionality, is the fact that ENCODE and similar projects have shown that 'the vast majority of the mammalian genome is differentially transcribed in precise cell-specific patterns'.[430] Indeed, it was this cell-type specificity that led

Ewan Birney to predict that 80 per cent functionality might be an underestimate, because of the expectation that, as greater numbers of cell types were studied than in the ENCODE findings published in 2012, then more regions of the genome were likely to show activity in these new cell types. It's important at this point, though, to mention a criticism that has been raised about some of the cells that ENCODE analysed. The project studied a large variety of different human cell types; the idea being to survey genome activity across the whole human body. But this meant working with many so-called 'immortalized cell lines'—cells that have acquired the ability to divide indefinitely, like a cancer. Such cells have played an important role in medical research ever since the first such cell line was isolated from a woman called Henrietta Lacks in 1951.[432] Lacks was a poor black woman from Maryland in the US who was admitted to Johns Hopkins Hospital after feeling a 'knot' inside her; in fact, it was a highly malignant type of cervical cancer. Johns Hopkins had a progressive policy of treating poor people for free. There was another less benevolent side to the institution though, for during Lacks' treatment, and without her knowledge or consent, a sample of her tumour was removed by Howard Jones, the doctor treating her, and given to George Otto Gey, a clinician who was experimenting with ways to grow human cells in culture.

Previous attempts to culture normal human cells had all failed. We now know this is because such cells can only divide a finite number of times, typically about fifty, before they die. This 'Hayflick limit', named after Leonard Hayflick, the scientist who discovered it in 1962, is thought to be one reason why we all have finite lives, although ageing involves many other factors as well. However, unlike all previous attempts, Henrietta Lacks' cells not only multiplied at a phenomenal rate in the culture dish, but kept on dividing.[432] Tragically, Lacks succumbed to her cancer soon after being admitted to hospital. Her cells, though, had found immortality, and continue to be propagated in laboratories across the world to this day, having been used to develop a polio vaccine, played an important role in research into cancer and AIDS, and been used to assess the effects of radiation and toxic substances on human cells. Gey called Lacks' cells HeLa cells in a clumsy attempt to protect her anonymity, and for many years they were believed to have come from someone called Helen Lane. Lacks' family only found out by accident how important the cells had become in 1973, when they were asked for a blood sample by a scientist studying the genetics of HeLa. And only in August 2013, after a long campaign for justice, did the family finally receive some acknowledgement for the unethical way they and Henrietta herself had been treated.[433]

In addition to showing that culture of cells outside the body was possible, the development of HeLa cells opened the way for the creation of many other immortalized human cell lines. While initially these were derived from cancers, just as HeLa cells had been, the development of ways to 'immortalize' cells in culture by exposing them to mutagenic chemicals, radiation, or adding oncogenes from viruses, made it possible to create a range of cell lines that retained properties of the organs or tissues from which they had originated. It was such cell lines that ENCODE used for some of its analyses, as well as stem cell lines that also have the capacity to divide indefinitely. Yet such types of cells are known to be transcriptionally 'permissive', making it possible that some of the high levels of transcriptional activity detected by ENCODE reflect this peculiarity, rather than being a general characteristic of all cells in the body.

The use of immortalized cell lines by ENCODE is one reason why we should treat some of its findings with caution. However, it's also important to note, as Stamatoyannopoulos does, that in contrast to the 'perception that ENCODE is largely a cell line centered endeavor ... overall ENCODE has sampled a vast range of primary cell types—indeed, these outnumber immortalized cell lines nearly three-to-one'.[434] Such 'primary' cells have been taken straight from the normal body, not immortalized. In addition, other studies have indicated that many functional elements identified by ENCODE, such as non-coding RNAs, show precise patterns of expression during embryo development and differentiation of stem cells into specialized cell types, as well as distinctive expression patterns across a complex structure like the brain.[435] Nevertheless, it remains possible that biochemical 'noise' might also have a precise expression pattern, which is why the ultimate way to test whether different elements in the genome are functional is to interfere with such elements in an experimental organism like a mouse, and assess the effect on the cell or organism.

The only problem with such a test is that, as we've seen, a surprising conclusion from mouse 'knockouts' of protein-coding genes is that elimination of the activity of a gene that has been identified as important by other criteria often has little effect on the whole organism, or very different effects to what was expected, presumably because of 'compensation' by other genes.[436] And if this is true of protein-coding genes, it's likely to be at least as much an issue for non-coding regions of the genome. As such, if knocking out such a region does not have a significant negative effect on the organism, this will not necessarily mean that it does not have an important function. Another problem in trying to test the function of non-coding regions of the genome is the sheer number of such elements. Using traditional methods of gene knockout this would have made it difficult to test elements on such a scale. However, recently new methods of 'gene editing' are dramatically reducing the cost and also rapidly speeding up the process of making knockout animals.[437] As such, it is becoming possible to test the function of non-protein-coding elements across the breadth of the genome in ways that would have been undreamt of only a decade ago.

Chris Ponting and his colleague, Andrew Bassett at Oxford University, have recently demonstrated that such gene editing can be used to rapidly test the function of non-coding RNAs on a large scale in model organisms like the mouse, but also the zebrafish; the latter being particularly important for studying vertebrate development because its embryos grow outside the mother and are transparent, making them very amenable for studying embryogenesis.[438] However, these scientists have also drawn attention to the potential pitfalls of such an approach, namely that assessment of the effects of such intervention requires a clear idea of which cell types and tissues are likely to be affected, based on analysis of the expression pattern of such non-coding RNAs. In addition, interventions in the genome need to be carefully designed so that there are no unintended effects upon protein-coding genes.[439] Such considerations are already an issue for studies of knockouts of protein-coding genes, since, in a number of cases, effects upon the whole organism have been misinterpreted because attempts to knock out one gene ended up affecting a neighbouring gene because of unintended disruption of DNA elements that regulated the latter.[440] Indeed, as we become increasingly aware of the complexity of gene expression, and the densely packed nature of functional elements in the genome, some of which may be superimposed upon each other, so care will be needed when interfering with such elements and interpreting the results of such interventions.

A different problem in assessing the functional significance of different regions of the human genome is the fact that ENCODE showed that a significant proportion of the biochemical activity detected in the genome was only found in humans, and not in the mice which were being used as a point of comparison. This finding is one reason why some critics of ENCODE are sceptical about the conclusions drawn by its researchers, since, if these regions really are functionally important, the findings would suggest that humans are far more different to mice than had been concluded by comparison of protein-coding sequences alone. To test whether this is the case, we, however, also face a practical problem, since if mouse knockouts are incapable of testing the functional significance of these genomic activities because they are only active in humans, not in mice, what alternative approaches will be available?

One possible route, if such regions of genomic activity are also found in other primates, would be to develop knockout versions of such species. Here the new gene editing technology is potentially of great importance, since one of the problems with traditional knockout technology was that it was effective in mice, but not other mammalian species. In contrast, the new technology can be applied to many species, with the result that the first knockout monkeys were recently created using such technology.[441] If regions of biochemical activity are identified that are only found in humans and other primates, but not in mice, for the first time it will now be possible to test the functional importance of such regions in a monkey model. However, if such regions are only found in our closest relatives, chimpanzees and other great apes, such an approach could pose ethical issues, given the opposition of some people to experimentation on such species, or indeed on any primate species.[441]

Of course, we face an even bigger problem in assessing the functional significance of biochemical activities in the genome that are restricted to humans alone. For, although such regions, if truly functional, might hold keys to the unique features that distinguish human beings from other species, how can we assess the significance of such regions given the impossibility, for ethical reasons, of creating genetically engineered knockout humans? One possible route would be to focus on human cells in culture. And, indeed, the new gene editing methods can easily be applied to human cells.[442] This means that it is now becoming almost routine to delete different regions of the human genome in cultured cells and then assess the effect on cell physiology. However, although this may allow characterization of the role of such regions in defined cellular processes, the approach has limited value for identifying important functions within a complex organ like the brain, or the body as a whole.

Critics of animal experimentation often argue that animal or human cells in culture could easily serve as a substitute for use of live animals in research. In fact, such studies of cells in culture already form a central part of a typical research project. To take my own studies of the roles of chemical signals in important physiological processes as an example, at least half our recent research has used cultured cells. However, while we can learn a lot about basic cellular processes by studying cells in a culture dish, to understand the role of chemical signals and the genomic processes that they control in their full complexity, it is necessary to study the whole organism. This is not only because complex organs, such as the brain or the heart, are impossible to grow in culture, but also because different organs communicate with each other via hormones and other signalling molecules in a way that is only possible to study in a whole, living animal.

Faced with the impossibility of genetically engineering human beings as part of experimental science, there is, however, a different route to identifying functional roles for the non-coding parts of our genomes by looking for links between these regions and human disease. As we saw in Chapter 6, one surprising conclusion of studies of the link between the genome and disease that have been taking place since the completion of the Human Genome Project is that at least 90 per cent of such links are not in protein-coding genes but in the rest of the genome. This finding is, in itself, an important piece of evidence in support of the idea that such non-coding regions of the genome are important, but it also provides a way to begin to assess the effect of naturally occurring mutations in the non-coding genome. In effect, the individuals identified in this way constitute natural 'knock-outs', or other types of mutants in which the function of a particular gene is not knocked out but altered in some way. Because of this, studying these individuals may, on the one hand, reveal insights about the underlying molecular mechanisms of their disease, but also provide important information about the role of the non-coding genome in normal bodily function. There is a catch, however: the genetic basis of human disease is itself turning out to be far more complicated than many people had predicted, as we'll now explore in Chapter 11.

11

GENES AND DISEASE

'The success of the Human Genome Project will also soon let us see the essences of mental disease. Only after we understand them at the genetic level can we rationally seek out appropriate therapies for such illnesses as schizophrenia and bipolar disease.'   *Jim Watson*

'We have, in truth, learned nothing from the genome other than probabilities. How does a one or three percent increased risk for something translate into the clinic? It is useless information.'   *Craig Venter*

Genes and disease have been inextricably linked ever since the birth of human genetics. Maybe this reflects a general tendency of people to particularly notice abnormal or curious characteristics rather than more commonplace ones in fellow humans. Indeed, as we saw in Chapter 1, the first human condition to be linked to a genetic mechanism at the turn of the twentieth century was alkaptonuria, a disease that drew the attention of Archibald Garrod because of its association with urine that turns black on contact with air.[443] Garrod's insights eventually opened the door to recognition of a succession of other human diseases that followed Mendel's laws (see Figure 21). So, conditions like Huntington's chorea and cystic fibrosis are dominant and recessive diseases, respectively. In addition, there are sex-linked disorders, caused by defects in the X and Y chromosomes.[444,445] Because the Y chromosome is so small and contains few genes, Y-linked disorders are rare and few in number.[444] X-linked recessive disorders generally only affect men, since women have two X chromosomes (although there are exceptions, such as women with a condition called Turner's syndrome, who have only one X chromosome).[445] However, women can pass on these disorders to their sons. Haemophilia, where sufferers can bleed to death from a minor injury because of a defect in the blood-clotting response, is an X-linked disorder. The condition has

152

CONDITION	GENE (CHR. LOCATION)	INHERITANCE PATTERN
Congenital Deafness (nonsyndromic)	Connexin 26 (13q11)	Recessive
Tay–Sachs	Hexosaminidase A (15q23)	Recessive
Familial hypercholesterolaemia	LDL receptor (19p13)	Dominant
Sickle cell anaemia	Beta-globin (11p15)	Recessive
Duchenne muscular dystrophy	Dystrophin (Xq21)	X-linked Recessive
Cystic Fibrosis	CFTR (7q31)	Recessive
Hemochromatosis	HFE (6p21)	Recessive
Huntington disease	Huntingtin (4p16)	Dominant

Figure 21. The different classes of single-gene disorders in humans

been known since the 2nd century AD, and ancient Jewish laws recognized that if a woman had two sons that died from circumcision her third son would not be required to be circumcised, showing some awareness of women as carriers.[446] Famously, Queen Victoria passed on the condition to many European royal males.[447] Indeed, the fact that the Russian Tsar and Tsarina's son had the disorder, and subsequently enlisted the monk and supposed faith healer Rasputin to treat the child, has been proposed as one of the destabilizing influences on the royal court that helped trigger the Russian Revolution. The discovery in 1991 of the remains of the Russian royal family, executed in 1918 at the height of the civil war that swept the country after the revolution, led to the demonstration that the haemophilia was due to a mutation in the intron–exon boundary of exon 4 of the clotting factor IX gene, showing this was a splicing disorder.[447]

The recognition that some human disorders followed the same inheritance patterns as Mendel's pea plants was a major step forward in human genetics. However, even the discovery that genes are made of DNA did not initially make it any easier to identify the specific gene defects responsible for such disorders, with the exception of conditions that affect the haemoglobin protein, like sickle cell anaemia and the thalassaemias. Here the link was obvious, given that these disorders were clearly connected with failure of the blood to transport oxygen in the normal manner. While both are recessive disorders, sickle cell is caused by a single amino acid change in the haemoglobin protein,[448] while thalassaemias are generally caused by a failure to properly produce the protein in normal amounts.[449] In 1949, Linus Pauling, who would receive a Nobel Prize in 1954 for his pioneering studies of protein structure, showed that the haemoglobin in sickle cell had an altered mobility upon separation using a technique called gel electrophoresis. This was the first demonstration of a link between an altered gene product and a disease, and Pauling predicted that 'medicine is just now entering into a new era [in which] scientists will have discovered the molecular basis of diseases, and will have discovered why molecules of certain drugs are effective in treatment, and others are not'.[450] However, these blood disorders are very much exceptions, and in general it has only been possible to identify the molecular defect responsible for single-gene disorders by a laborious search through the genome. This only became realizable in the 1970s when it became possible to cut and paste genes and sequence the DNA code itself. Armed with such tools, from the mid-1980s onwards, the molecular secrets of a series of single-gene disorders were finally revealed. So we now know the gene defects associated with recessive disorders like cystic fibrosis, dominant ones like Huntington's, and X-linked diseases like Duchenne's muscular dystrophy.

Huntington's has been recognized as a disorder since medieval times, but was only properly defined in 1872 when George Huntington described its successive symptoms of jerky movements, psychosis, and eventually full-scale dementia.[451] Identified as a dominant Mendelian disorder, the search for the affected gene was led by Nancy Wexler, who devoted her life to this quest after her own mother died of the disease.[452] A major breakthrough came with the discovery of an isolated community in Venezuela where a staggering half of the population were sufferers. Such concentrations of a specific genetic disease in a population are due to the 'founder effect', whereby a whole region is populated by descendants of an original carrier of a gene defect. This particular case provided an extended family of thousands of people who could be compared in genetic linkage studies, leading to the discovery of the huntingtin gene in 1993. This gene has a series of 'trinucleotide repeats', that is, the sequence CAG repeated over and over, at the start of its protein-coding region, resulting in a corresponding repeat of the amino acid glutamine.[451] Everyone has such repeats in their huntingtin protein; however, if an individual has more than thirty-five repeats, they will succumb to the disease, with the age at which they do so being related to how many extra repeats they have.

Huntington's disease has spread through the population in the past because most sufferers generally only show signs of the disorder between the ages of

35 and 44 years old. So people with the disease have often had children before realizing they themselves were sufferers, and, being a dominant disorder, the chance of these children succumbing is 50 per cent. One obstacle to treating the disorder is our continuing lack of understanding of the normal function of the huntingtin protein. Why the mutant protein causes the symptoms it does also remains unclear, although recent studies suggest that presence of the abnormal protein leads to enhanced cell death.[451] So, more than twenty years after the discovery of the huntingtin gene, the only really tangible outcome for sufferers is that a test is now available that shows how many CAG repeats an individual has in this gene, and therefore whether they will eventually succumb to the disease. A recent article in *The Guardian* newspaper by journalist Charlotte Raven, who took the test and now knows that she will eventually succumb to the disease, provides a moving account of what it feels like to live with this terrible knowledge.[453]

Much more is known about the proteins that are defective in cystic fibrosis and muscular dystrophy, recessive conditions associated not with the abnormal actions of a dominant mutant protein, but with the absence of a properly functioning normal protein. Cystic fibrosis has been known about at least since the eighteenth century, when literature warned 'woe to the child who tastes salty from a kiss on the brow, for he is cursed and soon must die', referring to the extra salt in the sweat of sufferers of this condition.[454] The disorder was only properly characterized in 1938, when Dorothy Andersen described patients with severe malfunction of the pancreas, the organ that produces our digestive enzymes as well as hormones like insulin, and linked such patients with those suffering from a lung disorder that left them highly vulnerable to asphyxiation and lung infections. In 1989, a team led by Francis Collins, who would later head the Human Genome Project, finally identified the defective gene, after a search which he compared to 'trying to find a burned-out light bulb in a house located somewhere between the East and West coasts without knowing the state, much less the town or street the house is on'.[455] The gene codes for the cystic fibrosis transporter protein, or CFTR, which regulates movement of chloride ions in and out of the 'epithelial' cells that form the inner boundaries of the lungs, pancreas, and some other tissues.[454] Lack of a functional CFTR protein leads to a build-up of salt in the sweat, and thick, sticky mucus accumulation in the lungs and pancreas, interfering with both digestion and breathing.

The most severe form of muscular dystrophy was first defined clinically in the mid-nineteenth century by Guillaume Duchenne, who gave his name to the disorder that he noted in boys who became progressively weaker, lost the ability to walk, and generally died in their teens.[456] Recently, a parent of a boy diagnosed with the disorder has described how what began as a routine check-up to find out why his son was slow in reaching certain pre-school 'milestones' turned into the realization that he would 'never play rugby, never make love, never make it to university, never realize his full potential'.[457] Duchenne's muscular dystrophy generally occurs in males and is carried by females, showing that it is an X-linked recessive disorder. In 1986, Louis Kunkel of Harvard Medical School identified the dystrophin gene, aided by the fact that some sufferers had obvious chunks missing from the X chromosome region where the gene was located.[456] In normal muscles the dystrophin protein links the surface membrane of the cell and its cytoskeleton, the protein structure that gives it shape and form. Lack of dystrophin destabilizes this interaction, leading to death of cells and muscle decay.[456]

When the CFTR gene was identified in 1989, the front cover of the journal *Science* featured a 4-year-old boy with cystic fibrosis sitting cross-legged framed by a rainbow of chromosomes.[458] Inside, editor Daniel Koshland confidently predicted that 'one in two thousand children born each year with a fatal defect now has a greater chance for a happy future'.[458] Geneticist Peter Goodfellow, who, a year later, would identify the SRY gene that single-handedly triggers the development of maleness in humans and other mammals, said 'the implications of this research are profound: there will be large spin offs in basic biology, especially in cell physiology, but the largest impact will be medical'.[458] However, we still lack a cure for these disorders despite having a much more detailed understanding of how the normal proteins work, and how defects in them lead to disease, so that Jack Riordan, co-discoverer of the CFTR gene, recently said that 'the disease has contributed much more to science than science has contributed to the disease'.[458]

One reason for this gap between understanding and ability to generate practical therapies is the difficulty in reconstituting a functional version of the proteins defective in these diseases, in the cells of a living person. Such 'gene therapy' is hampered both by the difficulties in getting artificial gene expression constructs into cells and expressing the missing protein without affecting expression of other genes. With Huntington's the goal is not to replace a missing functional gene, but to block expression of a dominant mutant version. As we've discussed, RNA interference may hold some promise for this type of disorder, since siRNAs can precisely target a specific RNA sequence and thus block the expression of a mutant protein, but not the normal version. Here also though, a primary obstacle is effective delivery of siRNAs to a cell. Maybe this is why, despite the existence of a genetic test for Huntington's, many people at risk through a known family connection choose not to take the test; ironically, this includes Nancy Wexler, who helped develop it. Wexler believes it has taken so long to find a cure because 'every time you look under a rock for what the Huntington gene's doing, you find something fascinating and interesting, maybe relevant and maybe not. And so even figuring out what's relevant is tricky.'[452]

Not that the clinical scenario for single-gene disorders is totally bleak. Phenyl-ketonuria, or PKU, is a recessive disorder affecting the enzyme that catalyses the transformation of the amino acid phenylalanine into tyrosine.[459] Because of this, phenylalanine entering the body as a component of many types of foodstuffs cannot be broken down, and instead accumulates to dangerous levels, leading to severe mental retardation, hyperactivity, and seizures during early childhood. Understanding the molecular nature of PKU has made a major difference to its treatment, by the simple practice of restricting phenylalanine in the diet.[460] Since excess phenylalanine only affects the developing brain, a more normal diet can be eaten after the teenage years. However, women with the disorder who become pregnant face a problem, since, although their children are unlikely to have the disorder because it is recessive, during the foetal stage they are highly vulnerable to their mother's blood. That is why pregnant women with PKU are advised to revert to an extremely strict diet, or not have children at all. PKU stands out as a success story partly because of the simplicity of the treatment. Unfortu-nately, many other metabolic disorders have proven less amenable to treatment simply by a change of diet.

As such, there is great interest still in the potential of gene therapy. The most promising situations are those in which the affected cells are most accessible, and indeed there has been partial success with a disease affecting a particularly accessible set of cells—those that make up the blood. One genetic disorder affecting white blood cells is an X-linked disorder, severe combined immunodefi-ciency, or SCID.[461] As we've seen, a key aspect of successful immunity is the production of antibodies by lymphocytes. This interaction is mediated by hor-mone-like substances called cytokines, and one type of SCID is caused by defects in genes coding for these cytokines.[461] Such individuals effectively have no immune system and are extremely vulnerable to infectious diseases. Because lymphocytes are produced in the bone marrow, SCID has been a popular target for gene therapy, since affected cells can be removed, treated, and put back into the body.[461] However, a major issue to be properly resolved is how to safely introduce a gene expression construct into bone marrow cells.

One strategy is to use retroviruses. As we saw in Chapter 8, retroviral RNA genomes are transformed into DNA by the enzyme reverse transcriptase and then insert themselves into the genome of the host cell. This is one reason why HIV can go unnoticed in a person's body for so many years; it also means that retroviruses offer a way to transport gene constructs into the genome of a target cell therapeutically. A clinical trial carried out in France in the late 1990s used retroviruses engineered to carry a cytokine gene into the bone marrow cells of boys suffering from SCID.[461] In one respect the trial was a great success, with 17 out of 20 boys regaining a functional immune system. However, a serious problem with this approach became evident when five of the boys developed leukaemia. It became clear that insertion of the retroviral DNA into the host cell genome had activated an oncogene, those genes that normally play important roles in cell growth but which can cause cancer if overstimulated. The high proportion of those affected suggested that the insertion next to an oncogene wasn't accidental, although why remains unclear. While the leukaemia was subsequently treated, the problems it highlighted led to the suspension of the trial. Despite this setback there is still great interest in developing safe forms of gene therapy. One hope is that retroviruses can be engineered to target the genome without disrupting expression of other genes, although, given the number of regulatory elements identified by ENCODE, this may be far from easy.[461] Alternatively, adenoviruses can deliver proteins to cells without disrupting the genome.

Despite these problems, there is no doubt that the identification of genes defective in diseases with a Mendelian pattern of inheritance has greatly advanced our understanding of the molecular basis of these diseases, and hopefully, therefore, the possibility of developing effective ways to treat them. However, even for single-gene disorders much remains unclear about their manifestation across a population. So, while some cystic fibrosis sufferers die in their teens from lung failure, others only realize they have two faulty CFTR genes when they present at the infertility clinic—the male reproductive system also being affected by this disorder—having no other obvious symptoms of the disease.[462] This variability is called the 'penetrance' of a disease. To some extent, it is due to different mutations in a gene having differing effects on the resulting protein; however, it also reflects the effects of other gene variants in an individual, and the unique environmental influences that individuals are subjected to during their lives, which can either counteract, or enhance, the effects of the gene variant in question.[463] So, sometimes the exact same gene mutation leads to severe symptoms in one person, but has no effect in another.[464]

If this is an issue for single-gene disorders, how about more common conditions such as heart disease, diabetes, cancer, and disorders of the mind like schizophrenia, bipolar disorder, and depression? A major selling point of the Human Genome Project was the claim that it would lead to greater understanding of such disorders. So, upon completion of the project in 2003, British science minister Lord Sainsbury said 'we now have the possibility of achieving all we ever hoped for from medicine'.[465] Daniel Koshland, editor of *Science*, promised that the basis for 'illnesses such as manic depression, Alzheimer's, schizophrenia and heart disease' would all be unravelled, with new drug treatments for these conditions sure to follow.[466] These pronouncements were echoed by Craig Venter, leader of the privately funded rival to the official genome project, who said, 'it is my belief that the basic knowledge that we're providing to the world will have a profound impact on the human condition and the treatments for disease and our view of our place on the biological continuum'.[467]

An important complement to the genome project was the creation of 'biobanks' of DNA samples and medical information.[468] For instance, the UK Biobank aims to collect such samples from half a million individuals. The idea is then to identify gene variants associated with specific diseases in these individuals. In contrast to previous approaches that tested the role in disease of pre-selected 'candidate' genes, this approach is presumed to be unbiased, covering, as it does, the whole genome. This is possible because of the existence of maps of genetic 'markers' that occur at points along each of the 23 human chromosomes. One class of markers are called 'single nucleotide polymorphisms' or SNPs, alterations in specific nucleotide bases that vary between different human individuals.[469] Although SNPs may themselves cause a disease, importantly, this need not be the case. Rather, SNPs can associate with the real genetic cause of a disease because they are close enough on the chromosome to be 'linked' together, during the crossing over that takes place during meiosis.

Genome-wide association studies, or GWAS, have been a major focus of biomedical research in the decade since the completion of the genome project.[470] Such studies have examined over 200 diseases and human conditions, with over 4,000 SNP associations made. One estimate is that at least \$250 million has been spent on such studies. So how successful has this approach been both in terms of helping us understand the molecular basis of diseases, and, as importantly, identifying new ways to diagnose and treat them? On the positive side, GWAS have highlighted some important novel links between genes and some common diseases, for instance, diabetes.[470,471] This condition, as commonly known, is associated with a high level of glucose in the blood; however, there are many other adverse effects, because insulin plays such a central role in the body. Insulin, which is secreted by the pancreas, is often known as the hormone of plenty because it regulates the accumulation of carbohydrate and fat stores and the growth of muscle following food intake. It exerts its effects by binding to a protein receptor on the surface of the cells it targets. This leads both to the transport of glucose into cells, but also regulation of a range of enzymes and other proteins involved in cell and tissue growth.

Type 1 diabetes occurs early in life and generally results in a total inability of the pancreas to produce insulin.[472] In contrast, type 2 diabetes occurs later in life and obesity is a major risk factor.[471] There is currently much talk about a diabetes 'epidemic', since the incidence of the disease has dramatically risen with rising levels of obesity. Since type 2 diabetes usually begins with an inability of tissues to respond to insulin, it was generally expected that defects would be concentrated in genes coding for the insulin receptor itself or other proteins that transmit its influence within the cell. Yet, GWAS have overwhelmingly identified a role in this disorder for genes involved in the formation and function of pancreatic beta cells.[470,471] This suggests that, although obesity is a trigger, individuals who succumb to type 2 diabetes are primarily those whose beta cells are less able to cope with the requirement for increased insulin production that occurs following the hormone's inability to manifest its effects in the body. In contrast, type 1 diabetes is mainly due to problems of auto-immunity.[472] Normally, the immune system distinguishes a person's own cells from bacteria, viruses, and other pathogens invading the body. However, this ability to distinguish self from non-self sometimes breaks down, resulting in auto-immunity. Type 1 diabetes is generally caused by auto-immune mechanisms destroying the pancreatic beta cells that secrete insulin, although, in rare cases, the defect is in a regulatory region of the insulin gene.[472]

Despite these insights, there have been growing concerns in recent years about the practical value of GWAS. Surprisingly, it's not that this approach has failed to find links between the specific region of the genome and disease, but rather that it has identified a bewildering number of such links—but their effects are predicted to be tiny. Recently, Sir Alec Jeffreys of Leicester University, discoverer of the genetic fingerprinting technique, has argued that 'one of the great hopes for GWAS was that, in the same way that huge numbers of Mendelian disorders were pinned down at the DNA level and the gene and mutations involved identified, it would be possible to simply extrapolate from single gene disorders to complex multigenic disorders. That really hasn't happened.'[473] Or, as Jon McClellan and Mary-Claire King of Washington University have noted, 'to date, genome-wide association studies (GWAS) have published hundreds of common variants whose . . . frequencies are statistically correlated with various illnesses and traits. However, the vast majority of such variants have no established biological relevance to disease or clinical utility for prognosis or treatment.'[474] Perhaps most damning for the argument that GWAS simply need to improve their methods or the number of people being analysed, McClellan and King conclude that 'the general failure to confirm common risk variants is not due to a failure to carry out GWAS properly. The problem is underlying biology, not the operationalization of study design.'[475] These various points are worth considering in detail. We mentioned in Chapter 6 how 90 per cent of links between common diseases and the genome are in the non-protein-coding regions we've been discussing in this book. Yet media reports of such findings still tend to have headlines such as one that appeared recently on the BBC website entitled 'Eighty new genes linked to schizophrenia',[476] despite the fact that the links identified are overwhelmingly not to genes as traditionally defined, but to these non-coding regions. Indeed, an important aspect of the ENCODE project was its demonstration that many GWAS links that were disregarded because they were not in protein-coding genes, map closely to areas of the genome associated with important gene 'switches', or obvious biochemical activity.[477]

This at least helps to address previous concerns that most GWAS 'hits' are meaningless in terms of their link with gene function. However, a more fundamental problem is revealed by the study highlighted by the BBC, in which more than 100 genetic regions were linked to schizophrenia, 83 of which had never been pinpointed before.[478] The leader of the study, Michael O'Donovan of Cardiff University, has concluded that 'finding a whole new bunch of genetic associations opens a window for well-informed experiments to unlock the biology of this condition and we hope ultimately new treatments'.[476] However, the fact that each of these one hundred or so differences is only predicted to have a tiny impact in terms of susceptibility to the disorder has led some critics to doubt how relevant such findings are for diagnosis and treatment.

The debate about the extent to which genetics determines susceptibility to mental disorders as compared to the influence of the environment is a long-standing one. The idea that mental illness is primarily a product of the environment reached its height in the work of psychiatrist R. D. Laing.[479] In his books *The Divided Self* and *Sanity, Madness and the Family*, Laing emphasized the pressures of modern life, particularly those within the nuclear family, as crucial triggers for schizophrenia; more controversially, he saw the disorder as a rational response to the 'madness' of modern society. Indeed, Laing once claimed that a schizophrenic teenager he encountered in a mental hospital who 'was terrified because the atom bomb was inside her', was less estranged from reality 'than the statesmen of the world who boast and threaten that they have Doomsday weapons'.[480] Influential during the 1960s, when mental illness could be viewed as part of a spectrum of rebelliousness personified by the phrase 'turn on, tune in, drop out', one problem with this viewpoint was that it ran the risk of glamorizing, and therefore potentially ignoring, the very real psychological pain and debilitating nature of schizophrenia for those who suffer from it.

In contrast, Jim Watson believes that schizophrenia is a straightforward genetic condition. Watson, whose own son Rufus has the condition, believes he has 'seen the failure of the environmental approach in a very personal way' since, 'for too long, my wife and I hoped that what Rufus needed was an appropriate challenge on which to focus. But as he passed into adolescence, I feared the origin of his diminished life lay in his genes. It was this realisation that led me to help to bring the human genome project into existence.'[481] However, the view that schizophrenia is purely a genetic disorder has trouble explaining some important facts. A recent study showed that black people of Caribbean origin living in Britain are nine times as likely to be diagnosed as schizophrenic as white Britons.[482] One explanation for this finding from a purely genetic point of view would be that this particular population is biologically more susceptible to schizophrenia; however, this fails to explain why incidence levels amongst black people in the Caribbean itself are similar to those of British whites. The study concluded that racism was probably a key factor, both in terms of being diagnosed schizophrenic and as a trigger of the condition, but also that other factors specific to particular immigrant communities, such as differences in family structure, may explain why Afro-Caribbeans are so vulnerable in this respect. But, whatever the exact reasons, the study suggests that that 'biological or genetic susceptibility do not appear to explain high rates of schizophrenia in black Caribbeans'.[483]

Another problem with the idea that schizophrenia is a specific 'disease' with a common biological origin is the sheer diversity of symptoms used to classify it. These range from 'delusions, hallucinations, loosening of associations, disorganized speech and behaviour, illogical thinking, social isolation and cognitive deficit'.[484] Confusingly, one individual can be classified as schizophrenic by having one set of symptoms while another has a completely different set. This suggests that, rather than being a single 'disease', schizophrenia may encompass multiple related disorders. This ambiguity is also true of other mental conditions such as bipolar disorder, depression, and autism. Indeed, recent studies have suggested that these different disorders share elements in common and may even have common causes; moreover, there is a considerable overlap between such conditions and many behaviours that are considered 'normal'.[485] Coupled with this lack of precision in terms of diagnosis is a lack of understanding about the biological basis of mental disorders. So, while some scientists view schizophrenia as a problem of brain development, others believe it is due to degeneration of nerve cells. A sensible viewpoint would, therefore, seem to be that social factors play an important role in the development of mental disorders, but susceptibility to conditions such as schizophrenia, which probably encompasses a range of different disorders, is affected by real biological differences between individuals.[485] However, if these differences are due to many genetic variables, each with a tiny effect, it is perhaps understandable that some have concluded, like Craig Venter, that this is 'useless information' for identification of new diagnoses and treatments, not just for schizophrenia but other common conditions where GWAS have revealed a large number of weakly contributing factors.[486]

Recently, however, this scenario has been challenged by another possibility, which is that, far from common diseases being a product of many genetic differences all having a small, but cumulative effect on the body, instead, they may be due to much rarer differences with a very powerful effect, but only in a few individuals.[487] The reason why this second scenario might be true is because of a potential flaw in GWAS, namely the assumption that common diseases are caused by common genetic differences that occur in at least 5 per cent of the population. This assumption has been partly a matter of necessity, since the databases that have been available do not include rarer genetic markers. However, it is possible that the links identified by GWAS actually reflect linkage with much rarer genetic differences that occur in a much smaller percentage of the population, these being the real causal agents of the disease.[487,488] A problem in testing such a possibility has been the prohibitive cost of sequencing whole genomes of many individuals in order to identify such rare differences. However, 'next-generation' approaches that generate DNA sequence for a fraction of the time and price of the Sanger method are now making it possible to address this issue directly.[489] And, tantalizingly, a recent study that focused on areas of the genome identified as linked to schizophrenia by GWAS, and carried out detailed sequence analysis in different individuals, did indeed identify rare differences that may have a considerable effect on specific genomic regions in certain individuals.[490]

A similar assessment as to which of the two scenarios just mentioned is correct is taking place in genetic studies of other common diseases.[471,488] In fact, as a recent review pointed out, 'it is likely that in a heterogeneous, complex genetic disorder such as schizophrenia, a subset of cases may be attributable to rare mutations with large effects while another subset may develop the disorder as a result of an interaction of multiple common variants of small effect'.[491] The exact situation will be important to resolve in terms of the future diagnosis and treatment of disease, for if a large number of common genetic differences contribute only a small amount to common diseases, this could make it very difficult to use such information to predict the chance of someone succumbing to a particular disease. Given that a number of commercial companies currently offer a disease prediction package based on known links between common SNPs and different diseases, this is an important concern.[492] The predictions made by such companies are advertised as giving individuals valuable information about their health and lifestyle choices. But they might also lead some people to worry unnecessarily about their health.

An original hope of GWAS was that the genes identified would be important new targets for drug design. While large numbers of potential targets have been identified, some critics have questioned whether a drug targeting a gene product that only accounts for a small part of a disease's genetic effect will be particularly effective therapeutically. If, however, only one or a few rare genetic differences are strongly linked to the disease, but differently in each individual, this could imply a much stronger causative role for genetics in common conditions.[492] Such a conclusion poses its own problems for diagnosis and treatment though, for it would suggest that there is no common genetic mechanism for susceptibility to common diseases. As such, it could provide hope for the idea of a personalized medicine for each individual, but would also pose practical problems for drug companies seeking drug targets that would translate into pharmaceuticals that could be used to treat the population as a whole.[492]

Although detailed analysis of the genomes of large numbers of individuals is now being carried out, economic considerations mean that this effort is still largely focused on the coding exons of genes, the so-called 'exome'.[487,488] However, as we've seen, 90 per cent of current genetic links with disease are outside the protein-coding regions,[493] so this approach may fail to identify the great majority of such links. Further new developments in sequencing will, therefore, be necessary if detailed analysis of the whole of the genome in many individuals is to become possible, thus establishing which of the two scenarios discussed is true. Whatever the outcome of this analysis, it's clear that the idea that a few genes in most individuals would determine susceptibility to common diseases now looks false. Another unresolved issue is why the combined influence of all the genetic variants identified as contributing to a particular disease by GWAS seems to be much less than that estimated from studies of disease incidence in families, especially comparisons between identical and non-identical twins.[494] There are a number of possible explanations for such 'missing heritability'. One is that many more genetic variants still remain to be identified. However, another possibility is that the influence of such variants and the environment upon a particular disease do not combine in the simple additive fashion that has been supposed, reflecting the fact that the interactions between different gene products in the cell, whether proteins or non-coding RNAs, is a highly complex affair.[494] But if this is true of genetic variants and disease, then what does it tell us about the link between the genome and human characteristics generally? It's time to investigate what features of our genomes might make us specifically human, and, at the same time, distinguish us from other individuals. But first we need to define what it means to be human at all.

12

WHAT MAKES US HUMAN?

'Man is descended from a hairy, tailed quadruped, probably arboreal in its habits.'
Charles Darwin

'The tool's function is to serve as the conductor of human influence on the object of activity; it is externally oriented...The sign, on the other hand...is aimed at mastering oneself.'
Lev Vygotsky

Skeleton of a crocodile. Nipple of a cat. Nose of a pig. Hair of a poodle. Thumb of a monkey. Eyes of a baboon. Brain of a chimpanzee. If this sounds like a list of ingredients for a witches' cauldron, think again, for it is merely a reminder of how many general characteristics we share with other animals. In *The Origin of Species* in 1859, Darwin was careful not to directly tackle the issue of human beings' relationship to other organisms on the planet, saying only, cryptically, that 'light will be thrown on the origin of man and his history'.[495] In fact, it was only in 1871, with *The Descent of Man*, that he applied his theory of natural selection directly to our own species. The book's major theme was that human beings share similarities in basic anatomy, physiology, and embryo development with other mammalian species. In *The Expression of the Emotions in Man and Animals*, published the following year, Darwin showed he wasn't afraid to include human behaviour and society in this comparison. In contrast, as we've seen, Alfred Wallace, co-discoverer of natural selection, found it unacceptable that this mechanism could produce such an amazing entity as the human mind. Instead, he argued that while natural selection could explain the workings of the human body, our consciousness must have some more supernatural origin. It was this that led Darwin to retort 'I hope you have not murdered too completely your own and my child.'[496]

In fact, Darwin's fears were groundless, for it is the materialist view that has triumphed. It is now a starting point in neuroscience and psychology that any understanding of human consciousness must be based on molecular and cellular mechanisms shared with other species.[497] This viewpoint was greatly strengthened by the discovery of the DNA code, since comparing protein-coding sequences between humans and other mammalian species indicates that we share a huge amount in common with such species. So, as we saw in Chapter 6, humans and chimps share 99 per cent DNA similarity in protein-coding genes,[498] and even the tiny mouse is 85 per cent similar in this respect.[499] These similarities not only confirm Darwin's view of a continuum between ourselves and other organisms, but also justify the use of animals as surrogates for humans in studies of the mechanisms underlying our biology.[500] In terms of basic bodily functions there is much that is attractive about this approach. The way the heart, lungs, or kidneys of a mouse work is very similar to how these organs operate in our own bodies.[500] Within the cell, the same chemical messengers, like cAMP, control key physiological processes, while in the blood and extracellular fluid, the same hormones, cytokines, and neurochemicals, regulate similar functions in all mammals. So, although falling in or out of love, savouring a fine meal or rueing a bad one, appreciating a great piece of music or regretting a poor one, may seem particularly human activities, the chemicals shaping these responses are essentially the same as those producing pleasure and pain in our pet cat or the mouse that it's pursuing.[501]

Yet the fact remains that what is most remarkable about human beings is not what we have in common with other animals, but what distinguishes us as a species. One obvious difference is the technologies that are such familiar elements of modern human life. So, to write these words, I am using a desktop computer, whose speed and memory would have been undreamt of only decades ago. From time to time I will use the Internet to research a point, check my e-mail, or engage in any of the distractions that comes with continual access to the World Wide Web. This is all done in a building made of bricks and mortar, fully supplied with water, gas, and electricity. Outside, I can hear cars or the occasional plane. Being surrounded by such technology it is easy for those in the developed world to forget just how remarkable is our current existence. But while millions around the world live in far more primitive conditions, they still employ a staggering variety of different technologies, whether an Amazon tribesperson with their bow and arrow, or an Inuit building an igloo from snow blocks in the Arctic.

Of course, human beings are not the only species to use tools to transform the natural world. Although tool use was initially thought to occur only in primates, studies have shown that some birds also use sticks as tools.[502] What distinguishes humans, though, is that our tool use continues to evolve. In contrast, there is little sign that the life of a chimpanzee in the jungle is particularly different in comparison to when our two species diverged 7 to 8 million years ago. Yet, in the last 50,000 years, humans have gone from living in caves to sending spaceships to Mars. Therefore, there is clearly something unique about the way we humans have made tool use a systematic part of our lives, and our capacity to invent new tools with each new generation.

Another unique attribute of our species that underlies these abilities is our capacity for language. Compared to the sounds that animals make, human language is unique in being a conceptual framework of symbols that allow us to describe things in the world, their properties and current location, but also what happened to them in the past and might happen in the future.[503] Humans seem to almost effortlessly learn this complex symbolic system if exposed to it at a sufficiently early age. Evidence for the importance of such a 'critical period' comes from cases of unfortunate individuals who, through abuse, have been deprived of human contact in their early years, and who can subsequently learn words but seem incapable of fitting them into a conceptual framework of the type just mentioned.[504]

Our understanding of the biological basis of this critical period, and the unique language-learning capabilities of human children, has been greatly boosted by new ways to safely and non-invasively image brain activity in babies and toddlers newly exposed to language. Commenting on these approaches, neuroscientist and linguist Patricia Kuhl of Washington University said recently that 'this decade may represent the dawn of a golden age with regard to the developmental neuroscience of language in humans'.[505] Such techniques, referred to by acronyms like EEG, MEG, NIRS, and fMRI, either measure electrical properties of the brain or its metabolism, and can reveal changes occurring as rapidly as milliseconds, as well as their position in the brain. Such studies have shown that the critical period is not a single 'window' but rather successive ones attuned to different aspects of language—sounds, syntax, vocabulary—and are helping to define areas of the brain involved in language learning.

If our ability to transform the world using tools and our language capabilities are central to being human, how did they arise in the first place? To address this issue, we need to delve deeper into what is known about human evolution. When Darwin wrote the *Descent of Man*, the fossil evidence for human evolution was so sparse that he had to guess the likely sequence of events that led to modern human beings.[506] In so doing, he was influenced by the mainstream religious and philosophical thinking of his time, which saw rational thought as the key motor of cultural change. Accordingly, while acknowledging the importance of bipedalism, Darwin proposed that the key difference between apes and the first proto-humans was the latter's possession of a large brain.[506] To explain why a large brain would have evolved, he suggested that it had been stimulated by the growth of language, initially as a warning system. Finally, using their large brains, our ancestors developed tools.

We now know that such a sequence of events is wrong. Surprisingly, the person who guessed the correct order was not a biologist, but Friedrich Engels, better known for his political writings and partnership with Karl Marx.[506] Engels was familiar with Darwin's account of human evolution in the *Descent of Man*, but became unconvinced by the order of events proposed there. Instead, in an essay written in 1876, called 'The Part Played by Labour in the Transition from Ape to Man', he argued that the evolution of the human brain was a consequence, not a cause of tool use.[507] Engels proposed that our ape-like ancestors began to stand upright, this then 'freed the hands' for using tools, which led to socialized labour and the new opportunities and demands this imposed on human society.[507] Finally, this led to the growth of the brain and, at the same time, language, for human beings 'now had something to say to each other'. Importantly, the growth of the brain and of language and culture then further stimulated the growth of the brain. Engels believed his conception was a novel one, asking Marx to keep quiet about the idea, 'so that no lousy Englishman may steal it'.[508] However, he never published the essay in his lifetime, and instead it only appeared in print in 1896 in *Die Neue Zeit*, a German socialist newspaper, a year after Engels' death. There, it was ignored by the scientific world, which was to have unfortunate consequences for our understanding of human evolution.[506]

So when a crude hoax consisting of a human cranium joined to an orangutan's jawbone was planted in an archaeological dig in Piltdown, England, in 1912, it was taken seriously for fifty years because it fitted Darwin's proposed sequence of human evolution. Meanwhile, Raymond Dart's discovery in South Africa in 1924 of the genuine partial remains of a creature with an ape-sized brain, but which was fully bipedal, was not taken seriously because it did not fit this sequence. In fact, it was only in the 1970s, when Donald Johanson found the famous 'Lucy', a complete 3.5 million-year-old skeleton of a small-brained bipedal ape, and Louis and Mary Leakey identified tool-using proto-humans with small brains, that the idea of bipedalism and tool use preceding the dramatic growth of the brain became accepted fact.[506]

While the development by proto-humans of different tools and their brain growth is now well established thanks to diverse fossil remains, the role that language played in our evolution remains much more uncertain.[509] This is because language leaves no trace in the fossil record, apart from indirectly in the structure of the mouths and throats of our ancestors. Difficulties in reaching a consensus about language evolution led the Linguistic Society of Paris in 1866 to ban debate on the topic because speculation was so far removed from any real evidence, a prohibition that influenced linguistics until the 1970s.[510] Since then, however, there have been more concerted attempts to study this question. To supplement the limited fossil evidence, studies have compared vocalization between ourselves and other species, investigated differences between existing human languages, and looked to indirect evidence for the existence of language in prehistoric societies such as signs of art, culture, and religion, which are presumed to require a complex language structure to sustain them.[511]

One suggestion put forward by Thomas Suddendorf of Queensland University, Australia, is that human language must have evolved in a society with significant levels of mutual trust.[512] The very separation of words as purely abstract symbols from the objects they relate to makes them ideal tools for deceit. However, while it is certainly possible to tell a bare-faced lie without batting an eyelid, it's precisely because our language ability is so conducive to lying that some believe it could only have arisen in a society operating on mutual trust.[512] This raises the question of why such a situation would arise in the first place. One possibility is that if human tool use evolved as a specifically social activity this would require a degree of trustworthiness for people to be able to work together. More positively, such shared labour would have led to the need to plan and organize such activities. The requirement for such planning could, in turn, have stimulated the development of a communication system that provided a sense of past, present, and future.

Another theory about human language, proposed by Robin Dunbar of Oxford University, sees it as having evolved to bind emerging human societies together.[513] In ape societies grooming helps to create social bonds and defuse tensions, and the 'gossip' theory of language origins sees it as a form of verbal grooming. However, it is important to recognize that human language is far more than a method of communication. Instead, it is tightly associated with a conceptual framework of symbols that makes our relationship to the world completely different to any other species, and any theory of its origins needs to explain this difference.[514]

So how did such a symbolic mind arise? A key challenge in answering this question is to explain how human consciousness, on the one hand, springs from the molecular and cellular mechanisms of an individual brain, and, on the other, is integrally linked to all the other 7 billion human brains on the planet by our location as individuals within wider human culture. One person who particularly sought to bridge this gap was the Russian psychologist Lev Vygotsky. He developed his theories about the mind over a decade, stretching from the early 1920s to his early death from tuberculosis in 1934, in a burst of creativity that led to some labelling him the 'Mozart of psychology' when his writings were rediscovered in the 1960s, having been banned by Stalin for their 'subversive' content.[515] Challenging the common view that human consciousness is primarily an individual affair, Vygotsky saw consciousness as a social construct, not in the crude sense of something written on a 'blank slate', but rather involving a complicated process of interaction between society and the inner psyche.[515] He drew a parallel between the way tools as external objects are 'aimed at mastering, and triumphing over, nature', with language based on words as inner objects, being 'a means of internal activity aimed at mastering oneself'.[516]

Vygotsky's studies of young children talking to themselves as they play showed that such 'egocentric speech' not only guides the child's activity but later becomes internalized as 'inner speech' and helps to create the thought processes of the individual. These studies also indicated that acquisition of speech restructures the brain so that thought is transformed from simple association to a process guided by a hierarchically ordered conceptual framework. This is an active process on the child's part, whereby the child seeks out the words and concepts that make sense of their everyday practical and social experience. Clearly, this ability must be based on real biological differences between humans and other species, but it also requires a social environment and interaction with other human individuals. Vygotsky also believed that mental disorders like schizophrenia are characterized by a partial or complete breakdown of conceptual thinking and a regression to the level of thinking by association.[517]

Having now sketched a picture of some essential features of what it means to be human, it is time to see how we might link such attributes to the human genome. Ultimately, our human uniqueness must be based on genetic differences between ourselves and other species. However, while there is much discussion about the link between genes and human behaviour and society in many popular accounts, how well does this relate to what we have been learning about the genome? In the Introduction to this book, I mentioned how claims are often made about 'genes' coding for complex human characteristics like nationalism, intelligence, personality, sexual persuasion, and even men's supposed unwillingness to do the ironing. I criticized such claims because, in general, no attempt was made to identify these proposed genes, or if such an attempt was made, as with the so-called 'gay gene', it ended in failure. In addition, posing the issue this way often reveals a naïve understanding of the complexities of human behaviour and society, and their historical nature. Further confirmation of the potential difficulties faced by those seeking an easy answer to the question of what role genetics plays in determining complex human characteristics was demonstrated by a recent study that sought to identify genetic differences linked to intelligence. Previous attempts to link IQ to specific variations in genomic DNA sequence 'have led to a slew of irreproducible results', according to a recent commentary in *Nature*.[518] In response, the biggest study yet to investigate the genetics of intelligence recently focused on more than 100,000 people, its aim being 'to bring more rigour to studies of how genes contribute to behaviour'. Yet this study identified only three variants associated with intelligence, and, according to the *Nature* commentary, 'their effects are maddeningly small', while, overall, the findings were 'inconclusive', meaning that 'scientists looking for the genes underlying intelligence are in for a slog'.[518]

So can we expect to find any intelligible genetic basis for the different facets of human behaviour and society? I believe that we can, but only by engaging with the new insights emerging about our genomes at the molecular level, coupled with a more sophisticated awareness of the complexity of the human body, and its interactions with the environment and with other human beings and the culture that we have created. A surprising aspect of many popular accounts is how their view of the gene seems anchored in the era of Mendel, when the gene was viewed as an abstract entity of no precise material form. Yet, as we've seen, not only do our genes' protein products exist in a cellular environment shared by thousands of other proteins, but the view of the genome as a linear entity upon which protein-coding genes are dotted like beads on a string, is being seriously challenged by new evidence that the genome is a complex 3D structure. In addition, the discovery that our genomes can respond to changes in both the cellular and the bodily environment through various epigenetic mechanisms means that the old division between nature and nurture appears far more fluid than previously suspected. At least in the lifetime of an individual, both the nutrients and also toxins our bodies encounter, as well as psychological stresses but also more positive experiences, may affect our genomes in a significant manner. What remains more uncertain and controversial is to what extent such epigenetic changes can be passed down to future generations. With these points in mind, it's time we looked more closely at what studies of our extinct ancestors, but also comparisons between humans and our closest animal relatives, can reveal about us both as humans and individuals.

A common but inaccurate view of evolution is as a ladder of progress in which species are like the rungs, with one species transforming into another in an orderly fashion. In contrast, Darwin himself saw evolution as more like a branching tree, with different species as offshoots.[519] Current theories of human evolution generally agree that *Homo sapiens* is merely the last existing member of a series of increasingly human-like species that diverged from our ape cousins about 6 million years ago. Studies of such past species have confirmed this progression, with ape-like creatures first beginning to walk upright (*Australopithecus afarensis*), use tools (*Homo habilis*), show signs of a rapidly growing brain (the unfortunately named *Homo erectus*), develop technology and perhaps even practise art in a more sophisticated manner (*Homo neanderthalensis*), eventually culminating in modern human beings (*Homo sapiens*) with all our unique attributes.[520] However, far from a ladder-like progression linking these species, there was considerable overlap in the period that they existed on the Earth, and this time was shared also with many other proto-human species (see Figure 22).

Fossil evidence has given us a sense of the physical forms of our proto-human ancestors: whether they walked upright and their height, the shape of their hands, the size and shape of their skulls, and, by extrapolation, their brains. Studies of

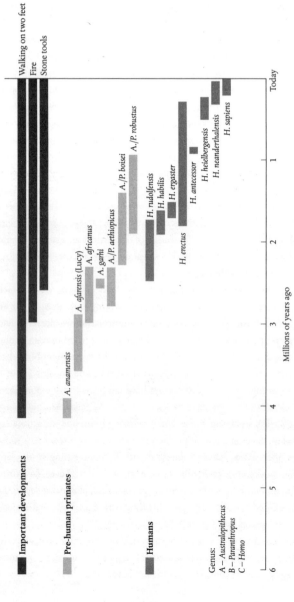

Figure 22. Human evolution timeline

tools and other artificial objects, works of art, and living spaces can all provide information about what might have gone on inside the heads of our ancestors.[521] But what of the genomes of these past species? Not so long ago it was thought impossible to address such a question directly. Recently, however, the Neanderthal genome and those of other proto-humans have been sequenced by Svante Pääbo and his team at the Max Planck Institute for Evolutionary Anthropology in Leipzig, making it possible to directly compare our own genomes with these extinct species.[522] Further understanding of how the human genome evolved has come from comparisons with our closest existing animal cousin, the chimpanzee, and by comparing human beings with each other.

In this latter respect, study of mitochondrial DNA has been very important. Mitochondria are the powerhouses of our cells, breaking down fats, sugars, and proteins, and transferring their energy to ATP, the cellular 'energy currency', in a process that requires oxygen.[523] The importance of this process for human life is shown by blocking it with cyanide, death following almost instantaneously. Yet mitochondria started off as free-living bacteria that became incorporated into our single-celled ancestors about 1.5 billion years ago, in a symbiotic relationship.[524] In line with this, mitochondria still retain some of their own genes, which have many similarities to those of bacteria. Mutations in mitochondrial genes cause defects in eyes, brain, heart, and muscle, these being high-energy-requiring tissues.[523,525] Such conditions are inherited from the mother, but affect both sexes; this pattern of transmission is due to the fact that, while egg and sperm both contain mitochondria, those of the sperm are destroyed once they enter the egg.[526]

Analysis of the mitochondrial genome in different human individuals has helped uncover the pattern of our evolution, since by estimating the mutation rate of the mitochondrial genome it is possible to treat the latter as a 'molecular clock'.[527] Two features of mitochondrial DNA—its maternal pattern of inheritance and the fact that, unlike chromosomal DNA, it's not reorganized by crossing over at each new generation—make it particularly amenable to studying the timescale of human origins, and the migration patterns of our ancestors. Insights have also come from studying the Y chromosome, since this is only inherited by males.[528] Such analysis suggests that *Homo sapiens* originated in East Africa about 150,000 years ago, from which point we eventually spread out across the world, first to the Middle East 60,000 years ago, and then to the rest of the world, reaching Europe 40,000 years ago.

Earlier groups of proto-humans, like *Homo erectus* and, later, *Homo neandertha-lensis*, also seem to have originated in Africa and then spread across the world. While *Homo erectus* appeared on the planet just under 2 million years ago and became extinct 150,000 years ago, Neanderthals appeared about a quarter of a million years ago and only became extinct 40,000 years ago. A question debated for many years is whether the emergence of our species led to the demise of the Neanderthals, and, if so, whether this was a direct or indirect effect. So, in William Golding's novel *The Inheritors*, written in 1955, we see prehistoric life from the perspective of a gentle, peaceful group of proto-humans, who meet a violent end at the hands of other proto-humans.[529] It is only at the end of the novel that we realize with a shock that the first, peaceful group are the Neanderthals, while their murderers are our own ancestors.

The idea of *Homo sapiens* wiping out other proto-human groups is certainly reconcilable with the fact that we have destroyed many other species, and continue to do so at alarming rates, since our advent on Earth.[530] However, the evidence for such a scenario is both sparse and contradictory, and it's just as likely that Neanderthals became extinct by simple out-competition for resources, or even because of changes in the climate and food resources that had nothing to do with modern humans.[531] Certainly, recent findings that indicate modern humans and Neanderthals coexisted in Europe for anything between 2,600 and 5,400 years, suggest that the process whereby Neanderthals disappeared once we met, was a gradual one.[532]

In addition, the picture has become complicated by the recent discovery that many people have a little bit of Neanderthal within them. Comparison of the Neanderthal genome with those of living human beings from across the world, shows that, on average, Europeans, but also people much further east in China and other parts of Asia, have around 1 to 3 per cent DNA of Neanderthal origin.[533] However, different people share different regions of their genomes with Neanderthals, so across the human population as a whole, humanity shares about 20 per cent of our genome with this species. In addition, another group of proto-humans, the Denisovans, ancient cousins of the Neanderthals whose remains were discovered in Siberia, share up to 8 per cent of their genome with modern Melanesian people who live in places such as Papua New Guinea and Fiji.[534] In contrast, people of pure sub-Saharan African descent have no Neanderthal or Denisovan DNA in their genomes, but may have residues of DNA from other proto-human groups that remained in Africa.[535]

Two recent detailed comparisons of Neanderthal and modern human genomes, one led by David Reich at Harvard Medical School, and the other by Joshua Akey at Washington University, showed that specific regions of the Neanderthal genome are represented but not others, suggesting that this selective retention has a functional basis.[536,537] Regions of the Neanderthal genome that are particularly prevalent in modern humans include genes active in keratinocytes, the cells that make skin, hair, and nails.[538] It is possible that acquiring these regions helped some humans to adapt to living in cold regions, since Neanderthals—who occupied a territory stretching from Western Europe to Siberia—were already adapted to a cold climate when they began to interbreed with modern humans newly arrived from the much hotter African subcontinent. A very specific adaptation of certain modern people living on the high altitude Tibetan plateau, the ability to thrive in a low oxygen environment, was recently shown to be due to a variant of a gene called EPAS1 that originated in Denisovans.[539]

We are also uncovering important clues about our own evolution from regions of the Neanderthal genome that are absent in modern humans. So, according to Akey, 'we find these gigantic holes in the human genomes where there are no surviving Neanderthal lineages. Most of these variations were removed in a couple of dozen generations.'[540] This suggests that such parts of the genome were harmful to human–Neanderthal hybrids and their descendants, and were purged rapidly as a consequence. That different living human individuals share different regions of their genomes with proto-humans, raises the question of whether different groups of people in the world, and individuals within those groups, have characteristics linked to these genomic regions. However, against such a possibility providing a biological basis to the view—typically used to assert white superiority—that humanity is strictly divided into different 'races', humans are far more similar genetically to each other than chimps, despite there being 7 billion of us compared with a few hundred thousand of them.[541]

An important reason for comparing the human genome with that of Neanderthals or other proto-humans is that this might allow us to identify genetic differences that underlie our unique attributes as a species. However, findings so far indicate that such differences are likely to be subtle. This is not surprising. Labelling someone a 'Neanderthal' usually signifies that a person is crude and uncultured; however, recent studies have challenged the view that our proto-human cousins were simple-minded brutes. There is evidence that Neanderthals consciously buried their dead,[542] and were aware of the beneficial properties of herbs—with one study suggesting they may even have been partial to a soothing brew of camomile tea.[543] They also made a kind of glue for securing spear points by heating birch sap while protecting it from air by a method so sophisticated that archaeologists have had trouble replicating it. Such findings have led João Zilhão, an archaeologist at Barcelona University, to argue that Neanderthals were capable of abstract thinking, just like modern humans, on the basis that 'burying your dead is symbolic behaviour. Making sophisticated chemical compounds in order to haft your stone tools implies a capacity to think in abstract ways, a capacity to plan ahead, that's fundamentally similar to ours.'[544] A hotly debated question is whether Neanderthals practised art. Many excavations of places where they lived have unearthed lumps of pigment—red ochre and black manganese—that were sometimes worn down as if they had served as prehistoric crayons.[544] In another site three cockleshells were found with holes near one edge, and traces of pigment, implying they might have been worn as ornaments. Zilhão believes this shows Neanderthals decorated themselves both with body paint and jewellery. Most controversial of all is the discovery of cave paintings in Spain that may substantially predate the arrival of modern humans in Europe. One problem here is the uncertainty about the precise age of these paintings, with different methods for determining their age giving different results. So, while one estimate is of around 40,000 years old, the period in which modern humans are thought to have reached this region, the paintings may be substantially older. Much hangs on the precise age, for the later date could either mean that the pictures were painted by modern humans, or that Neanderthals merely copied them.[544]

Such disagreements matter because they impact on the sort of differences we might expect to see in a comparison between our genomes and those of Neanderthals. Of the few obvious genetic differences so far detected between ourselves and our Neanderthal cousins besides differences in skin cells, one noticeable change is in genes linked to autism and schizophrenia. Autism is the name for a range of conditions associated with difficulties in socializing and communicating with other people, and a tendency towards stereotyped or repetitive behaviours. As we've seen, recent genome-wide association studies have undermined the idea that one or a few gene defects determine mental disorders. Nevertheless, some genetic links appear stronger than others, such as the autism susceptibility candidate 2, or AUTS2, gene.[545] First identified in identical twins that were both autistic, and subsequently in 36 unrelated individuals with autism and associated learning disabilities, many changes that affect this gene are in non-coding regions linked to its regulation. Recent studies in zebrafish and mice indicate that this gene codes for a transcription factor controlling the expression of genes involved in brain development. It is therefore interesting that AUTS2 and its associated regulatory regions were identified in a genome comparison between modern humans and Neanderthals as the most notable area of difference between the two species.[545] Other genomic regions that differ between ourselves and Neanderthals have also been linked to disorders of social interaction and learning.[546]

Further confirmation of a link between a tendency towards mental disorder and being human have come from a fascinating recent study that looked at epigenetic marks in Neanderthal and Denisovan genomes compared to those in modern humans.[547] As we saw in Chapter 6, chemical changes, like methylation in regulatory regions of genomic DNA, have a big impact upon whether a particular gene is turned on or off. It might seem an impossible task to identify such marks in proto-human genomes, given their extreme age and state of degradation, but researchers led by Liran Carmel at the University of Jerusalem, working with Pääbo's group in Leipzig, developed an ingenious way to do so by virtue of the fact that methylated nucleotides degrade to a different product than unmethylated ones do, during a process called deamination. This allowed the scientists to indirectly create a methylation map of such ancient genomes for the first time.[547] Since only two individuals—a Neanderthal and a Denisovan, both female—were analysed, the findings need to be treated as very preliminary, especially since the epigenomes of different individuals may vary considerably; however, Chris Stringer of London's Natural History Museum believes the study shows 'how we can begin to unlock epigenetic aspects of ancient genomes which have been hidden from us up to now'.[548] As well as finding differences in the activity profile of genes involved in skeletal development, which could account for the Neanderthal's shorter, stockier form and barrel chest, the study found that methylation differences between modern humans and Neanderthals were particularly prominent in genes linked to some mental disorders, with the suggestion that such genes were expressed at a much lower level in Neanderthals.[549]

These findings suggest that the very genetic changes that underlie modern humans' unique mental capabilities may also predispose us to such disorders. This further confirms the idea that attempts to find a gene 'for' schizophrenia or autism may be misplaced, not only because of the large number of different genomic regions now linked to these disorders, but because such regions may normally play integral roles in the distinctive mental processes that define us as humans, such as creativity, abstract thought, and capacity for complex language. This possibility would be in line with suggestions that disorders like autism and schizophrenia are part of a broad spectrum of states of mind that overlap with those of the normal population.[550] It would also fit with the idea that there is a thin line between genius and insanity, and explain why many gifted mathematicians and abstract thinkers have shown autistic characteristics. Indeed, Fredrik Ullén, of the Karolinska Institute in Stockholm, has recently found that both highly creative people and those suffering from schizophrenia have a lower density of D2 receptors that bind to the neurotransmitter dopamine in the thalamus area of the brain.[551] Noting that 'fewer D2 receptors in the thalamus probably means a lower degree of signal filtering, and thus a higher flow of information from the thalamus', Ullén believes that such a barrage of uncensored information may help fan the creative spark.[552] However, while allowing creative individuals to make unusual connections in problem-solving situations that other people might miss, distortions in this ability in schizophrenics could lead to disturbing and destabilizing thoughts.

Despite these steps forward in studying ancient genomes both in terms of their DNA sequence and even their methylation state, we've seen in previous chapters how a complete picture of genomic activity needs to encompass many more factors, such as the mRNA transcripts generated, the gene regulatory proteins that regulate transcription, the multiple types of histone modifications, and the noncoding RNAs that we now know regulate gene expression at a variety of different levels.[553] Such factors need to be studied in different cell types, and, as importantly, during the development and growth of the body. As such, and given the absence of any living proto-human species to conduct such studies on, an alternative way of investigating how these different factors all come together to make us human is to study them in different cell types and stages of development, in both humans and our closest living relatives, chimpanzees. And, in particular, such analysis has focused upon one particular organ—the brain.

13

THE GENOME THAT
BECAME CONSCIOUS

'An important stage of human thought will have been reached when the...
psychological, the objective and the subjective, are actually united, when the
tormenting conflicts or contradictions between my consciousness and my
body will have been factually resolved or discarded.'   *Ivan Pavlov*

'Science's biggest mystery is the nature of consciousness. It is not that we
possess bad or imperfect theories of human awareness; we simply have no
such theories at all.'   *Nick Herbert*

The human brain is the most complex structure in the known universe. Comparisons of the brain to a computer, or references to it as our 'wetware', analogous
to computer software, barely do justice to the true complexity of this organ. So
not only do our brains contain around 100 billion nerve cells, but each of these
can be connected to as many as 10,000 others, giving a total of some 100 trillion
nerve connections. As such, using our brains to try and understand how human
self-conscious awareness arose within this organ, why it is lacking in the brains
of our closest biological cousins, and relating this to the differences between
the human and chimp genomes, is undoubtedly the biggest challenge in biology
today. One of the most noticeable differences between humans and chimps is the
much greater brain/body ratio in our species. If it were down to differences in
body size alone, our brains should be 50 per cent bigger than those chimps; in
fact, they are 200 per cent larger.[554] We should be wary, though, of seeing size
alone as the defining feature of the human brain, since Neanderthals had an even
bigger brain/body ratio than ourselves.[555] Equally important is the fact that some
regions of the human brain are differently proportioned compared to chimps.

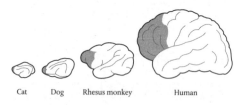

Figure 23. Prefrontal cortex in humans compared to other mammalian species

One region, called the prefrontal cortex, which has been implicated in complex thought, expression of personality, decision-making, and social interaction, is particularly pronounced in humans (see Figure 23).[554] Another region, the arcuate fasciculus, which connects parts of the brain implicated in language, is also visibly different in humans.

To gain insights into functional differences between our brains and that of chimps, recent studies have used a variety of different approaches. So Daniel Geschwind and his team at the University of California have been studying different patterns of expression of mRNA transcripts in human and chimp brains.[556] Because the brain is such a complex organ, such analysis has been carried out on numerous different brain regions (see Figure 24). This revealed that there are major differences in the levels of thousands of RNA transcripts between the two species. Surprisingly, few of these differences appear to be specific to a particular brain region. There are two possible explanations for this. One is that the unique attributes of the human brain evolved without large-scale changes in the gross functional and structural composition of these regions. If true, this would tend to go against the idea that human brains are compartmentalized into modules, each responsible for a different behaviour. This view of the brain, vividly captured by Harvard University linguist Steven Pinker's analogy with a Swiss army knife with its multiple gadgets, is in line with the fact that injuries to the brain can often result in some apparently quite specific defects, for instance in language ability.[557] The idea of a 'modular' brain was subsequently linked to the proposal that different human characteristics are coded by specific genes that are only expressed within these modules.

This view of the human brain and the behaviour that results from it was challenged by the discovery that our genomes contain just over 20,000 genes, not much more than a worm or a fruitfly. These findings make it difficult to see

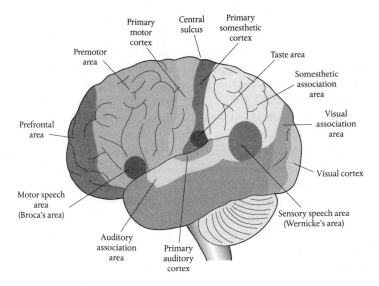

Figure 24. The human brain

how such a small number of genes could carry out the bewildering array of different human behaviours. As we saw in Chapters 4 and 5, though, genes have multiple ways to express themselves through alternative splicing, cell-type specific enhancers, and non-coding RNAs, so there is a danger of overemphasizing this point. The lack of differences in mRNA expression in different brain regions also argues against modularity, although it is possible that the regions analysed were too big to reveal differences in a few cell types. Finer-scale analysis will be required to address this issue. Fortunately, this is becoming increasingly feasible as new techniques make possible the study of the 'transcriptome' of single cells.

However, there are other reasons for doubting that particular human behaviours can be confined to specific regions of the brain in such a localized fashion. For instance, a recent study of electrical activity in the brains of monkeys engaged in problem-solving show that synchronization of brainwaves to form new communication circuits occurs across the prefrontal cortex and the striatum, two completely different brain regions.[558] Earl Miller of the Massachusetts Institute of Technology, who led the study, believes this shows 'there is some unknown mechanism that allows these resonance patterns to form, and these circuits start humming together'.[559] Miller thinks the findings demonstrate a division of labour between the two brain regions, so that 'the striatum learns the pieces of the puzzle, and then the prefrontal cortex puts the pieces of the puzzle together', and have relevance for understanding how the human mind can absorb and analyse new information.[559] Learning and long-term memory formation are known to be associated with changes in the brain's synapses, or connections between nerve cells. But Miller thinks this process of 'synaptic plasticity' is too slow to account for the flexibility of the human mind. As he points out, 'if you can change your thoughts from moment to moment, you can't be doing it by constantly making new connections and breaking them apart in your brain. Plasticity doesn't happen on that kind of time scale.' Instead, he believes that the synchronized 'humming' he has identified 'foster[s] subsequent long-term plasticity changes in the brain, so real anatomical circuits can form. But the first thing that happens is they start humming together.'[559] What remains to be shown is how such global waves of electricity in the brain are subsequently translated into the changes in gene expression that underlie synaptic plasticity. The most likely mechanism for connecting these two phenomena is via 'second messengers'—small molecules such as cAMP, nitric oxide, or even the humble calcium ion—which, as we saw in Chapter 3, play a central role in switching genes on or off. Being small, second messengers not only operate inside cells, but they can also traverse the whole brain because of pores called 'gap junctions' that connect the different cells of this organ together. So studies using chemical probes that fluoresce when they come into contact with calcium ions, such as those designed by Roger Tsien,[560] show that learning is accompanied by calcium 'signals' that are distributed across different regions of the brain, and show complex properties in both time and space.[561] Such signals can have an immediate effect by acting upon calcium-sensitive enzymes, but they can also trigger changes in gene expression by activating transcription factors via addition of phosphate groups by calcium-regulated kinases such as CaMKII. In fact, recent studies have identified CaMKII as a key player in learning and memory, while defects in the activity of this protein seem to be one cause of psychiatric disorders such as schizophrenia, depression, and epilepsy.[562] An important question for future studies will be to study how the spread of electrical and chemical signals across the brain differs in humans compared to other species, how this relates to differences in gene expression in different brain regions, and whether changes in these dynamic features of the human brain provide insights into some brain disorders.[561]

While insights can be gained by studying the differences between adult human and chimp brains, another important source of human uniqueness may lie in differences in brain development. As such, recent studies have investigated differences in mRNA levels in different regions of the brain at different developmental stages, both before and after birth.[554] One conclusion of such studies is that there are major differences in the times at which different genes are expressed during development in humans, compared to chimps, with some genes expressed substantially later in humans, others much earlier. Such differences are especially prominent in the prefrontal cortex compared to other brain regions. Genevieve Konopka, who studied this issue with Daniel Geschwind at the University of California, believes this shows that 'the intricate signalling pathways and enhanced cellular function that arose within the frontal lobe created a bridge to human evolution'.[563] Of particular interest is that genes involved in the formation of new synapses peak in expression several months after birth in chimp prefrontal cortex but only after 5 years of age in this region in humans (see Figure 25).[554] Moreover, electron microscope analysis of the synapses themselves shows that these are still being formed as late as 10 years old in humans.

According to Konopka, 'the biggest differences occurred in the expression of human genes involved in plasticity—the ability of the brain to process information and adapt'.[563] This would fit with the fact that human children have a much

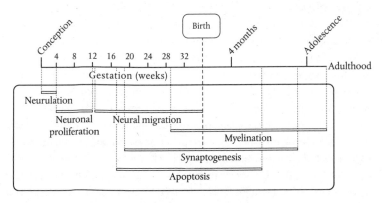

Figure 25. Timeline of human brain development

more extended period of learning than apes, and indicate that our brains are being restructured in response to such learning for a far longer period. Another important difference between humans and apes is the time at which 'myelination' of brain nerve cells takes place. The myelin sheath is a fatty, insulating layer that protects nerve cells and enhances the speed at which they transmit electrical impulses.[564] However, this comes at the expense of a capacity to form new connections and undergo changes in structure. Studies have shown that, while myelination is complete by the onset of puberty in chimps, in humans it still occurs up to the fourth decade of life.[554] Findings like these provide important new evidence for the idea that 'neoteny'—the retention of juvenile features in the adults of a species—has played an important role in human evolution. Evolutionary biologist Stephen Jay Gould, of Harvard University, in particular stressed the importance of this process on the basis that human adults have many physical features—a flatter and broader face, hairless body, large head to body ratio—in common with young, but not older, apes.[565] Such slowing of development was thought to allow a greater capacity for learning, but only now has its importance for brain evolution been confirmed at the molecular level.[554] Intriguingly, while we share many physical features in common with Neanderthals, their teeth seem to have matured much more rapidly than human teeth, akin to those of an ape, suggesting that, at least in some respects, the extent of neoteny in this species was less advanced than in modern humans.[566]

What would be interesting to explore is whether the restructuring of thought that Vygotsky believed took place in children via the 'tool' of language, and which we discussed in Chapter 12, can be linked to such dynamic changes in brain activity and subsequent changes in gene expression and plasticity. In this respect, an interesting finding of recent imaging studies is that there seem to be important overlaps between the human brain regions that mediate tool use and those governing language; however, these overlaps were not seen in other primates.[567] Another difference between humans and other primates is that human brains seem uniquely attuned to learning new tool use and language skills. This ability seems to be particularly associated with specific types of nerve cells called 'mirror' neurons, which can become activated both when a person is carrying out an action and when they are observing that action. This may explain one particular unique feature of infant humans, which is that they seem particularly capable of mimicking new actions and words; this could be one explanation why humans seem uniquely capable of both learning from past ways of working, as well as inventing new ones.

The thousands of differences identified in the patterns of gene expression in the human brain compared to the chimp suggest that the 1 per cent difference between ourselves and apes at the level of protein-coding genes may mask much greater differences in terms of functional gene expression. But this also complicates the process of trying to identify the key differences that led to the unique attributes of consciousness and self-awareness that characterize our species. As such, there is increasing interest in identifying whether these large-scale changes in gene expression are due to a much smaller number of 'master-controller' genes. And, indeed, by comparing the regulatory elements adjacent to the genes whose expression is different in humans, it has been possible to identify certain transcription factors that activate gene expression on a global scale. One such factor, called MEF2A, has generated interest not only because it is a master regulator of synapse formation, but also because one of the enhancers that controls its expression is mutated in humans but not in Neanderthals, suggesting this may have contributed to a delay in synapse formation in humans, but not in our extinct cousins.[568] Another transcription factor that controls the expression of many genes in the brain and is altered in humans compared to chimps is coded by the CLOCK gene. This is one of a class of genes that regulates the body's circadian rhythm, or body clock, which governs our sleep/wake cycle. However, according to Daniel Geschwind, recent findings suggest 'that it orchestrates another function essential to the human brain', most likely one linked to brain plasticity.[563] Disruptions in the action of this gene have been implicated in mania-like behaviour in humans, providing further evidence that genetic attributes that make us uniquely human may also make us susceptible to mental disorders.[569]

However, it is not only transcription factors that control gene expression; there is growing evidence of key roles for miRNAs. Recent studies have increasingly focused on identifying miRNAs with important functions in the human brain. One such miRNA, miR-184, which is abundant in the prefrontal cortex of human brains but not those of chimps, has previously been shown to be an important regulator of nerve stem cell proliferation.[570] This is interesting since neurogenesis—the growth of new neurons from such stem cells—is important not just during brain development in the embryo, but is also increasingly being recognized as a key process in the adult human brain.[571] Reflecting on this new-found role for non-coding RNAs in brain function, John Mattick and his colleague, Guy Barry of the University of Queensland in Australia, have recently argued that, while 'proteins form the core of basic cellular functioning...the increased sophistication, complexity, and plasticity of the regulatory RNA superstructure...has been at the heart of human cognitive advance'.[572] As evidence for this claim, Mattick and Barry point to studies showing that genomic regions that generate non-coding RNAs have been one of the main targets for mutation since humans diverged from chimps. They believe that emerging roles for numerous different classes of non-coding RNAs in the human brain suggest that they are 'temporally and spatially regulated to control both feedback ("hard-wired") processes during development and feed-forward ("soft-wired") processes during post-developmental cellular function'; this suggests that non-coding RNAs mediate processes like learning and memory through their link to multiple layers of further epigenetic processes.[572] One such process is RNA editing, in which RNA sequences are changed by conversion of nucleotide bases into modified forms. Interestingly, 'most of the edited sites occur in noncoding regions, implying that editing is not only modifying the structure-function properties of neuronal proteins, but also RNA-based regulatory circuits'.[573] That RNA editing in the brain is enhanced 35-fold in humans compared to mice, and has increased further during the transition from apes to humans, is seen by Mattick and Barry as further evidence of its important role in our evolution.

One important difference between the regulation of gene expression by non-coding RNAs compared to transcription factors, is that while the latter only control production of mRNAs, non-coding RNAs can affect the process at multiple levels. This has particular implications in nerve cells because of the highly differentiated structure of this cell type (see Figure 26). A typical nerve cell receives inputs of information from other nerve cells via structures called dendrites, of which it may have as many as 100,000. Its output to other nerve cells is concentrated on to a single axon; however, this usually has many branches, meaning that it can send signals to thousands of other nerve cells, but also to muscles and glands. The differing spatial reaches of nerve cells are shown by the fact that axons can be as short as a millimetre, or as long as a metre in the case of those that span the length of the spinal cord. When gene expression was thought to be regulated only at the level of mRNA production, it was assumed that all the

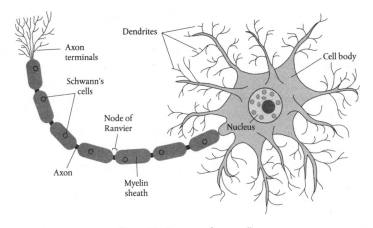

Figure 26. Structure of nerve cells

different parts of the nerve cell would contain similar types of proteins. However, it is now becoming clear that different regions of the nerve cell contain different populations of non-coding RNAs.[574] Given the capacity of non-coding RNAs to regulate translation of mRNAs into proteins, this suggests that specific dendrites or axon branches may have a unique protein profile, with important implications given the role of such structures in learning and memory. Moreover, the fact that one of the proteins whose translation is regulated by non-coding RNAs is the transcription factor CREB, whose role in learning and memory we discussed in Chapter 3, shows the potential complexity of the feedback processes involved.[575]

Such a focus on non-coding RNAs is important, given their rapidly emerging roles, but we should not forget the potential importance of other epigenetic mechanisms for human brain function, for instance, those involving chemical modifications to the DNA or the proteins that associate with it as part of the complex nature of chromatin. We've already discussed how epigenetic changes to the genome constitute a key way in which the cells of a developing embryo 'differentiate' into the myriad different cell types of the body, despite having identical genomes at the level of DNA sequence alone.[576] Such changes underlie the very different functional properties of such cells, for instance, the beat of a heart cell, the ability to conduct electrical impulses of a nerve cell, or the capacity of a liver cell to metabolize food and toxins. This differentiation of cells during development is triggered by changes in the extracellular environment, primarily by growth factors and 'morphogens'—chemicals that induce changes in cell form and function—but also by contacts between different cells, that trigger changes within the embryonic cell via receptors on its surface.[577] Such a change in the internal state of a cell in response to environmental signals also occurs during learning and memory in the brain. A surprising revelation in recent years has been the discovery of how similar differentiation and development are to memory and learning at the molecular level. So, a remarkable number of the same genes and signalling pathways that regulate differentiation and development are employed in forming the synapses that mediate learning and memory. In addition, there is increasing evidence of great similarity in the epigenetic mechanisms underlying these two different processes.

In Chapter 6 we discussed the concept of a histone 'code'.[578] Far from being a simple on/off switch, the incredible variety of different ways of modifying histones means that this type of regulation can be specific to one or a few genes, as well as being finely graded like a dimmer switch, and it even has a built-in timing mechanism due to the fact that different modifications can be reversed at very different rates. In addition, it appears that DNA methylation may also form a similarly sophisticated code.[578] All this makes this type of gene regulation potentially ideally suited to forming one of the molecular bases of the complex processes of learning and memory. Indeed, the far-sighted Francis Crick proposed something along these lines as early as 1984, when he suggested that 'memory might be coded in alternations to particular stretches of chromosomal DNA'.[579] What he lacked was a mechanism for such changes; however, a number of recent studies have shown that interfering with histone modifications and DNA methylation has profound effects on memory and learning processes in animal models.[580,581] What still remains to be shown is whether distinct changes in modified histones or the methyl state of particular nucleotides within specific genes are responsible for particular memories. Hopefully, such is the sophistication with which it is becoming possible to identify such changes in specific genes, this idea may soon be tested in an animal model of learning. In addition, increasingly refined genetic engineering techniques are making it feasible to interfere with specific epigenetic changes in mice, with a view to seeing how they regulate learning and memory.

Finally, as we saw in Chapter 8, an exciting development of recent years has been the suggestion that transposition events may play a role in the development of individual human personalities. In particular, the hippocampus, the brain region associated with memory formation but also with forming new nerve cells through 'neurogenesis', seems to be particularly prone to transposon activity.[582] This activity seems to be very sensitive to changes in the environment, being boosted, for instance, in mice undergoing exercise regimes compared to those which remained sedentary. This finding has led to the proposal that, in humans, increased transposition activity could either change behaviour, allowing the individual to become more adaptable to a new environment, or, alternatively, increase the risk of mental disorders, depending on the particular environmental pressure. Findings such as these challenge the long-held idea that, at the cellular and genomic level, all cells in an individual's brain are essentially equivalent, as well as the notion that all humans are broadly similar in their cellular and genomic properties.[582] Instead, an individual's life experience may profoundly affect the way their brain works, so that even identical twins, traditionally assumed to be genetically identical, may differ considerably in their brain operation, depending on their particular experiences in life.

Combining what we've learned from recent studies of the differences in gene expression in humans compared to chimps and proto-humans like Neanderthals, with new findings from the fossil record, there would appear to be two distinct phases in the evolution of the human brain.[554] The first phase saw a long and gradual increase in brain size, accompanied by important changes in certain brain regions, and was shared to a varying degree by other proto-humans. This was followed by a second phase about 150,000 years ago, which saw more subtle developmental remodelling of the brain, leading to the unique characteristics of human consciousness that define our particular species. It is this that presumably underlies the explosion around 50,000 years ago of new technologies, art, and culture in human society that is still ongoing, taking our species to every corner of the Earth and even into outer space.[583]

One recurring feature of many genes and their regulators that seem to play important roles in human consciousness is their link with mental disorders.[554] So why have such differences arisen if they leave our species vulnerable to such disorders? One suggestion is that the very speed of the changes that accompanied the evolution of human consciousness allowed insufficient time to fine-tune these new brain processes and render them robust enough to withstand environmental and natural genetic perturbations. Linking this possibility to their claims regarding the importance of non-coding RNAs in human brain function, Mattick and Barry have suggested that 'although the increase in mammalian cognitive ability has provided unique mechanisms to evolve exceptional skills, such as reasoning and awareness, it would also seem likely that a relatively new and increasingly complex regulatory system would have weaknesses and be vulnerable to stressors'.[584]

Another issue that remains to be properly addressed is how human evolution was able to occur so rapidly in the first place. As we saw in Chapter 1, the standard model of evolution is that change occurs through the natural selection of particular variants in a population, by virtue of their ability to survive, and, most importantly, reproduce, in a specific environment. That such different variants exist was discussed in Chapter 2 as being due to changes in the DNA sequence of the genome that occur because of the effects of radiation, chemicals, or other environmental insults. Such a mechanism is generally used to explain how our species arose, with initial mutations occurring that led to variants of apes that could walk upright, while subsequent mutations affected the ability to use tools, growth of the brain, and so on. However, a potential problem with this model of human evolution is whether it is rapid enough to account for the astounding speed in which our species developed from brute animals to self-conscious beings. This is particularly an issue for the most recent phase of human evolution, since, while it took several million years for proto-humans to evolve from apes, development of sophisticated technologies, art, and culture, which are assumed to have required a new kind of self-conscious awareness, only seem to have taken off as recently as 50,000 years ago, this event itself only occurring 100,000 years after the appearance of modern humans on the planet.[583]

It is with this in mind, but also because of increasing awareness of the importance of epigenetic mechanisms of gene regulation, which might allow more rapid forms of evolution, that questions are now being asked as to whether such mechanisms might also have played a role in recent human evolution, particularly of the brain. Such a possibility requires two things to be true: firstly, that epigenetic changes can occur in the genomes of nerve cells in the brain, and, secondly, that such changes can be passed down to future generations. However, while epigenetic changes and transposition events in the brain are becoming increasingly implicated as mediators of our behavioural responses to life's stresses, challenges, and opportunities, what evidence is there that such processes have had any long-term impact upon the evolution of our species?

As we mentioned in Chapter 2, a long-standing dogma, first put forward by August Weismann, is that whatever happens throughout our lifetimes to the non-sex cells in our bodies—the so-called 'somatic cells'—has no influence on future generations since it is only the sex cells, the eggs, and sperm, that pass on their genomes to our offspring.[585] What we also saw though, in Chapter 9, is that this dogma may finally be starting to be challenged. So, not only is there evidence from both animal experiments and observations of human populations that epigenetic changes may be passed down through several generations, but the mechanisms underlying such inheritance are now being identified. In particular, the finding that effects of stress in mice can be passed on to offspring by a route involving miRNAs in the sperm, and the detection of these miRNAs in the blood raises the possibility that the connection between the sex cells and the rest of the body, particularly the brain, may be more fluid than previously thought.[586]

Because of its capacity for radically restructuring the genome in a short space of time, transposition has been proposed as an important factor in human evolution. We saw in Chapter 8 how transposon activity can lead to disease by disrupting vital protein-coding genes. The potential threat to the continuation of a species, were such activity to occur freely in the eggs and sperm, has led to evolution of protective mechanisms, such as tight regulation of transposition by piRNAs. Yet we've also discussed a more creative aspect to transposon activity in the creation of new DNA regulatory elements. One puzzling aspect of transposition is that although increasing evidence suggests that this may have important functional roles in our brain cells, and therefore could conceivably increase our survival and reproduction prospects, this raises the question of what benefits the transposons themselves gain from their activity. More generally, could epigenetic events in the brain impact upon what is transmitted to future generations? In an interesting parallel, piRNAs, which were initially thought to be only expressed in the gonads, have now been shown to be also present in brain cells, and it has been suggested that they may play important roles in regulating specific patterns of transposition in individual nerve cells.[575] Could there be a more direct connection between the two organs than previously suspected? And, if so, could transposon activity in the brain affect the propagation of transposition effects to future offspring through the sperm, and therefore a transfer of the transposons themselves to the next generation? As yet, no mechanism has been identified for such a connection but, analogous to how stress can be passed on to subsequent generations via miRNAs, it seems likely that these non-coding RNAs would play a role in such a process.

In considering what sort of information might be passed down the generations through epigenetic means, the main focus has been upon the transmission of adverse effects such as stress. But there are also tantalizing hints that more positive life events might not only have a beneficial effect upon the brain through an epigenetic route in a person's lifetime, but that such effects might also influence further generations.[587] Quite how significant such effects are remains a matter of some controversy, and it is important at this point to introduce a note of caution, for, as Edith Heard of the Curie Institute and Robert Martienssen of the École Normale Supérieure in Paris, have warned, 'although the inheritance of epigenetic characters can certainly occur—particularly in plants—how much is due to the environment and the extent to which it happens in humans remain unclear'.[588] Heard and Martienssen point out that, while there does seem to be a basis for epigenetic influences extending across one or two generations in humans, as just outlined, evidence for more than this limited extent remains to be established. In addition, amidst the current interest in epigenetic mechanisms there is a danger in ignoring a far more obvious way in humans in which the experience of one generation affects those of the future. This is the fact, pointed out by Stephen Jay Gould, that 'human cultural evolution, in strong opposition to our biological history, is Lamarckian in character. What we learn in one generation, we transmit directly by teaching and writing.'[589] Of course, it's possible that both epigenetic mechanisms and social evolution might interact. If so, it seems likely that some of the 98 per cent of the genome previously assumed to be junk will play an important role in this process. Finally, if we have learned anything from recent discoveries about genomic and epigenetic mechanisms, it is that it's wise to keep an open mind. Or, as epigenetic researcher Brian Dias of Emory University, Atlanta, puts it, 'if science has taught me anything, it is to not discount the myriad ways of becoming and being'.[590] All of which means that, for those interested in the complex nexus of biological and social influence, even more exciting findings than those described in this book undoubtedly lie ahead.

CONCLUSION

The Case for Complexity

'The ultimate aim of the modern movement in biology is to explain all biology in terms of physics and chemistry.'   *Francis Crick*

'If we want to attain a living understanding of nature, we must become as flexible and mobile as nature herself.'   *Johann Wolfgang von Goethe*

From its inception, biology in the modern age has been characterized by a tension between two opposite poles. On the one hand, is the view expressed by seventeenth-century philosopher Francis Bacon that 'the nature of everything is best seen in its smallest portions',[591] a statement echoed by Francis Crick's claim that 'the ultimate aim of the modern movement in biology is to explain all biology in terms of physics and chemistry'.[592] On the other hand, there is the belief that biological systems have their own complex properties that must be explained in their own terms. First associated with the Romantic movement of the early nineteenth century, particularly through the utterances of poets like Coleridge and Wordsworth, this viewpoint was also held by some notable biologists at this time, such as Alexander von Humboldt and Lamarck, the latter expressing it through his belief that 'living beings have specific characteristics which cannot be reduced to those possessed by physical bodies'.[593] Indeed, these two activities could be combined in a single individual: so while Goethe is now primarily known as a literary figure, he also studied optics and the morphology of plants.

Goethe and his contemporaries faced a central problem, however, which was that the methods available to study the complexity that they recognized in nature were far too simple to do justice to it. So Lamarck sensed the potentially complex nature of the relationship between organism, environment, and inheritance, but was unable to explain in material terms what might mediate this relationship. And although Goethe himself believed that future generations would remember him for his scientific investigations rather than his literary works, the opposite has been true. In contrast, it is the very reductionist approach that the Romantic poets and scientists abhorred that has been the dominant trend for the last two centuries.

A particularly powerful aspect of reductionism in modern biology has been its capacity to focus on one or a few elements that are isolated and studied separately as a way to illuminate the whole. In this book we've seen this process at work from Darwin onwards, whose theory of natural selection seemed so simple and straightforward to Thomas Huxley that he famously remarked 'How extremely stupid not to have thought of that!'[594] In choosing to focus on competition for scarce resources as the primary element in the evolutionary process, Darwin and Wallace gave less emphasis to the importance of cooperation in nature. Yet it could be argued that this has played an equal role in the origin of our species, whether through symbiosis, as when our single-celled ancestors fused with a bacterium, the latter becoming an energy-providing mitochondrion in return for a sheltering environment, or the way that cooperation in tool-using apes set them on the path to language and human consciousness. Darwin himself was far from ignorant of such considerations, for instance, once stating that 'in the long history of humankind (and animal kind, too) those who learned to collaborate and improvise most effectively have prevailed'.[595] Instead, it was individuals like Herbert Spencer, who first coined the phrase 'survival of the fittest', who were more responsible for the emphasis on crude competition in popularizations of Darwinism.[596] Nevertheless, Darwin's primary focus on competition helped ensure the success of natural selection as a principle, but has also been a factor in the distorted presentation, in some popular accounts, of how evolution works.

In genetics we've seen a similar demonstration of the power, but also the potentially distorting effects, of simplifying a complex process. The genius of Mendel was to recognize simple mathematical patterns in complex chains of inheritance. This focus on mathematics may be one reason why the importance of his work went unrecognized in his own lifetime, since biologists at that time were unused to thinking in such quantitative terms. However, another reason was that when Mendel tried to convince Carl von Nägeli, one of the world's most distinguished botanists at the time, of the importance of his findings, he failed to do so partly because he couldn't demonstrate the same simple patterns of inheritance in the hawkweed plants that Nägeli sent him to test.[597] In fact, Mendel knew that even in peas some characteristics did not show the same straightforward rules of inheritance as the ones he used to illustrate his theory. Morgan also identified characteristics in fruitflies that failed to show a simple Mendelian pattern of inheritance but chose to ignore them.[598] Such simplification of genetics helped its success as an explanatory principle, and provided an important tool for understanding diseases like cystic fibrosis or Huntington's disease; however, it has also led to naïve expectations about the link between the genome and more common disorders that are only now being challenged by the reality of GWAS findings.

Another central principle in modern genetics is Weismann's proposal of a rigid division in multicellular organisms between the sex cells—the eggs and sperm—and the rest of the body, and his view that whatever happens in life to the body as a whole has no impact upon the only immortal part of the organism, the genetic material passed on to its descendants.[599] To back up his proposal, Weismann chose the rather barbaric route of cutting off the tails of 68 mice and showing that this did not result in any offspring born without tails over the next five generations.[599] In so doing, he helped boost the evolutionary synthesis of Darwinism and Mendelism; at the same time, his proposal seemed to shut the door firmly on Lamarck's view that the life experience of an organism could affect future generations, something that, as we've seen, is now being challenged in a number of important ways.

One view of Francis Crick's 'central dogma' of molecular biology is to see it as a modern version of Weismann's proposal.[600] So, just as Weismann's germ plasm was supposed to be the one pure and unchanging element of life, so, according to Richard Lewontin of Harvard University, DNA conceived as the central repository of information has assumed the mystical, self-renewing powers of the Holy Grail.[601] In cutting through the complexity of life as viewed by the biochemists, such a viewpoint played an invaluable role in opening up the genome to full view. Yet, at the same time, the one-sidedness of the central dogma has been exposed by the discovery of the myriad number of different ways in which DNA's information capacity can be modified by epigenetic mechanisms, both within the lifetime of an organism in different cells and tissues, and, more controversially, across generations too.

In summarizing some of the crucial developments of modern genetics, we can see both the strengths of the reductionist approach but also its weaknesses.

Typically, this approach seeks to reduce the complexity of life by focusing on one or two elements and then following these along a 'path of least resistance', just as a traveller may follow a path through a wood. There is no doubt that following such a path has allowed phenomenal progress in our understanding of genetics. However, just as a traveller will reach a destination by following a straight path, but may consequently miss important things on either side of the path, this, to some extent, has also been the case with molecular biology. Thankfully, though, this is not the end of the matter, because science also has an inbuilt mechanism that forces researchers to consider new paths of investigation even if this is not their original intention. We have seen this countless times in this book, where preconceptions about our genome, for instance, that it is a compact entity like that of a bacterium, were challenged by the discovery of enhancers, splicing, and so on. But it has also been true of the biggest biology project of all—the Human Genome Project.

Compared to the overwhelmingly positive headlines that greeted the 'first draft' of the genome, announced at a White House press conference in June 2000, media commentary about the tenth anniversary of this event was more subdued, and more critical. So a typical example was the *New York Times*, which, on 13 June 2010, ran a front-page story entitled 'A Decade Later, Gene Map Yields Few New Cures'.[602] Other media outlets and blogs across the world took up the theme that the project which had cost so much and taken almost a decade to complete had been largely an exercise in hype, and a failure in medical terms. One scientist particularly upset by such coverage was Eric Lander, director of the Broad Institute, a biomedical research institute in Cambridge, Massachusetts, and one of the leaders of the genome project. So he has rejected claims that he ever unduly hyped the project, arguing that 'I'm on record saying this is going to take a long time, and that the next step is to find the basis of disease, and then you have to make drugs...Going from the germ theory of disease to antibiotics that saved people's lives took 60 years. We might beat that. But anybody who thought in the year 2000 that we'd see cures in 2010 was smoking something.'[602]

In fact a survey of the claims made by those involved in the genome project at its inception uncovers some that are realistic, others that now seem overly simplistic, such as Jim Watson's claim in 1989 that 'we used to think our fate was in the stars. Now we know, in large measure, our fate is in our genes.'[603] Of course, one reason for such flowery phrases may have been the need to raise the $3 billion required to sequence the genome, which, in the 1980s, was an unprecedented amount of money for a science project. A vision was required to enthuse the government agencies that would be bankrolling the project, even if that vision now, in retrospect, looks unrealistic. But this and other pronouncements made at the time, for instance, about the links between genetics and disease that would be uncovered by the genome project, also seem to me to betray a genuine naïvety about the complexity of this link, rather than being simply an attempt to hoodwink politicians into handing over taxpayers' money. Yet against those who argue that the Human Genome Project was a waste of money, it is important to stress both the very real successes of the post-genomic age, and also the way in which the more unexpected aspects of the project's findings have subsequently led to a more sophisticated understanding of the genome's complexity, and how we might harness this understanding in practical ways.

Discussing progress in science, Sydney Brenner once said that this 'depends on new techniques, new discoveries and new ideas, probably in that order'.[604] In this book, we've discussed some remarkable technologies that have been brought to bear upon the question of how our genomes function, both to regulate intracellular processes but also act as a repository of information for the next generation. One distinctive feature of molecular biology is the fact that many of its tools are themselves derived from life, whether this be DNA or RNA probes, antibodies, fluorescent proteins, enzymes to cut and paste DNA, or the polymerase used in DNA sequencing. However, we should also not forget the X-ray diffraction devices that first revealed the structures of DNA and proteins, or the microscopes and fluorescence imaging instruments that allow us to peer inside a cell and even visualize the very DNA and protein molecules themselves. Perhaps the most important legacy of the Human Genome Project besides the genome sequence itself was the introduction of massive computing power into biology, as well as robotic devices that have led to 'high-throughput' approaches becoming a routine aspect of modern biology.[605] One important consequence is that DNA sequencing itself has been transformed: Sanger's original method now having been superseded by others based on massive numbers of reactions all proceeding in parallel.[606] It is this that lies behind the dramatic reduction in cost of sequencing an individual human genome over recent years, which, according to George Church of Harvard University, 'dropped by a factor of 10 every year for the last five to six years, so it's a truly amazing exponential decrease compared to the computer industry'.[602]

This reduction in costs has exciting implications for medicine. As we've seen, the original assumption that common disorders such as heart disease, diabetes, or disorders of the mind like schizophrenia or autism would turn out to be caused by defects in one or a few genes has been undermined by the findings of genome-wide association studies. Yet it would be equally mistaken to underestimate the increasing capacity of genomic studies to pinpoint the molecular causes of some diseases now that is becoming possible to sequence an individual genome both rapidly and cheaply. Take, for instance, a recent case in which a 2-month-old baby boy was admitted to Children's Mercy Hospital in Kansas City, USA, with a mysterious ailment that had already caused his liver to fail and left him hovering at death's door.[607] By sequencing the boy's genome, within three days geneticist Stephen Kingsmore and his team at the hospital had pinpointed the cause of the ailment to a mutation linked to a rare condition in which an overactive immune system damages the liver and spleen. Armed with this diagnosis, the boy's doctors immediately gave him drugs to lower his immune response, with the consequence that he is now at home and healthy. In fact, this baby is just one of 44 sick infants whose genomes Kingsmore's group has sequenced, using a process that can provide a diagnosis in as little as 24 hours. In 28 of these cases the researchers were able to diagnose the illness, and in half of these they could recommend a treatment. Over the next five years Kingsmore is planning to sequence the genomes of 500 sick babies at the hospital, and this is just one of a number of projects across the breadth of the USA that are waiting for approval to carry out a similar exercise. Of course, such projects raise a number of ethical issues. The babies in these cases have given no consent for their genomes to be sequenced. And while the focus in this analysis has been on identifying the molecular defects linked to specific, potentially life-threatening conditions, important questions will need to be asked about who will have access to the sequence data, and how far doctors should be allowed to proceed in extracting information that is unrelated to the immediate disease from which the children are suffering.[607]

Advances in DNA sequencing are just one example of the ways in which high-throughput methods are transforming biology. This now makes it possible to investigate the cell's activities on a truly global scale, whether the object of study is RNA transcripts, DNA methylation, histone marks, or 3D interactions. Indeed, this was how ENCODE mapped a diverse number of biochemical activities across the whole genome. And, as we've seen, the findings of such analyses are revealing an undreamt of complexity in the genome. Importantly, far from our biology being focused solely around the 2 per cent of our genomes that code for proteins, it is becoming clear that this is only the tip of the iceberg once we start to consider the 3D structure of the genome, non-coding RNAs, and all the epigenetic mechanisms that we've discussed.

As we've discussed in this book, a major part of the debate about the ENCODE findings has focused upon the question of what proportion of the genome is functional. Given that the two sides in this debate use quite different criteria to assess functionality it is likely that it will be some time before we have a clearer idea about who is most correct in this debate. Yet, in framing the debate in this quantitative way, there is a danger that we might lose sight of an exciting qualitative shift that has been taking place in biology over the last decade or so. So a previous emphasis on a linear flow of information, from DNA to RNA to protein through a genetic code, is now giving way to a much more complex picture in which multiple codes are superimposed upon each other. Such a viewpoint sees the gene as more than just a protein-coding unit; instead it can equally be seen as an accumulation of chemical modifications in the DNA or its associated histones, a site for non-coding RNA synthesis, or a nexus in a 3D network. Moreover, since we now know that multiple sites in the genome outside the protein-coding regions can produce RNAs, and that even many pseudo-genes are turning out to be functional, the very question of what constitutes a gene is now being challenged. Or, as Ed Weiss at the University of Pennsylvania recently put it, 'the concept of a gene is shredding'.[608] Such is the nature of the shift that now we face the challenge of not just recognizing the true scale of this complexity, but explaining how it all comes together to make a living, functioning, human being.

It's here, though, that we face a dilemma, which is whether the conceptual tools available to us in modern biology are sufficient to make sense of all this complexity and relate it not just to human disease, but also the other characteristics that distinguish us as a species and as individuals. In particular, although it is quite clear that reductionist methods have proven incredibly successful at identifying the molecules that make up the living cell and organism at an exquisite level of detail, what is now being debated is whether reductionism as a philosophy is capable of showing how all these molecules work together within an interrelated whole. Importantly, while such concerns go back at least as far as Goethe and his contemporaries, they are now starting to be raised within the heart of molecular biology itself.

Take, for instance, a recent article by Marc van Regenmortel of the École Normale Supérieure de Biotechnologie in Strasbourg, written for *EMBO Reports*, a journal of the European Molecular Biology Organization, in which he has argued that 'the reductionist method of dissecting biological systems into their constituent parts has been effective in explaining the chemical basis of numerous living processes. However, many biologists now realize that this approach has reached its limit.'[609] Van Regenmortel believes this is due to the fact that reductionism's assumption that 'the isolated molecules and their structure have sufficient explanatory power to provide an understanding of the whole system' does not take into account that 'biological systems are extremely complex and have emergent properties that cannot be explained, or even predicted, by studying their individual parts'.[609]

Of course, it is one thing to state this problem and quite another to find a way to make sense of such complexity. In this respect, some would argue that the global nature of the new technologies that have emerged following the genome project addresses this issue. So, as well as genomics, we now have approaches to catalogue all proteins (the proteome), RNA molecules (the transcriptome), metabolites (the metabolome), and interactions (the interactome).[602] But it is one thing to catalogue this complexity, another to understand its functional significance. In fact Sydney Brenner, who, as we saw in Chapter 2, played a key role in cracking the genetic code, amongst other major contributions to molecular biology, has recently argued that 'this "omic" science has corrupted us. It has created the idea that if you just collect a lot of data, it will all work out.'[602] Instead, Brenner believes that the way forward lies in the recognition that the organizing principle for thinking about the genome is the cell. So, in an article published in 2010, he outlined a project called CellMap, whose aim would be to catalogue every type of cell in the body and detail how different genetic regions behave in each cellular environment.[602] In fact, as we saw in Chapter 6, such an aim has been central to ENCODE, with its survey of genomic activity in 147 different cell types.

There is a potential problem, however, in exchanging a catalogue of molecules for one of cells, with regard to a structure as complicated as the human brain. For, as we've seen, far from the brain being a collection of isolated cells, we must consider the fact that each cell can have as many as 10,000 connections to others in the brain. One problem with merely cataloguing all the different types of brain cells and their genomic activities, is that this takes no account of how connections with other nerves might be affecting these activities. In fact, there are plans to address this issue, most notably through the Human Connectome Project.[610] This will map every single one of the 100 trillion nerve connections in the human brain by using sophisticated imaging techniques to study living brains, and high-throughput electron microscopy to study sections of dead brains. Amongst the thousands of individuals to be studied, the project will collect information from identical twins and their non-twin siblings, as well as individuals who suffer from various mental disorders. By mapping brain connections in these different cases, the project hopes to uncover the normal variation in human connectomes and how they change as humans learn, mature, and age.[610] Such is the aim, but the scale of the project is a daunting one. The connectome of the nematode worm, with 300 nerve cells joined by 7,000 connections, took a decade to complete. More recently, Hongkui Zeng and colleagues at the Allen Institute for Brain Science published a preliminary map of the connectome of the mouse brain, which has 75 million neurons. In comparison, the human brain has 100 billion nerve cells, as many as there are stars in the Milky Way. And, as Sebastian Seung, a neuroscientist at the Massachusetts Institute of Technology has pointed out, 'your connectome [has] a million times more connections than your genome has letters. Genomes are child's play compared with connectomes.'[610]

Such are the technical challenges faced by the connectome project, but there is also the question of whether the project really has the potential to deliver 'nothing less than the nature of human individuality', as is being promised in some quarters.[610] In particular, there are two potential flaws in the idea that simply mapping all the connections in the brain will be enough to reveal the underlying basis of human consciousness. One is the idea that an individual nerve cell can be treated as a 'black box' into which information is fed, and from which it emerges, without reference to the structure of the cell. Yet as we saw in Chapter 13, we now know that each nerve cell is composed of many functional sub-domains, with protein production in each regulated differently, depending on which type of non-coding RNAs it contains.[611] A second potential flaw is the focus on one-to-one connections between nerve cells. While, in itself, mapping all such connections in the brain is likely to be highly illuminating, it does not take into account the rapid and much more global connectivity that we've already discussed in the form of electrical impulses and changes in the levels of second messengers such as calcium ions, which can spread across the brain in seconds.[612] Since calcium signals can both regulate the activity of important enzymes, as well as gene expression, such global connectivity also needs to be taken into consideration. In fact, moves are already underway to address the second of these issues, with some neuroscientists calling for a further brain scanning project, the 'dynome', whose aim will be to go beyond mapping mere brain anatomy and instead connect this to an understanding of brain dynamics.[613] As for the issue of the many sub-domains within nerve cells, one hope is that, as the non-coding RNAs present within the domains of different types of nerve cells are studied, some kinds of generalized patterns will emerge, relating to the different sorts of such RNAs and their spatial distribution within different categories of nerve cells.

In addition, though, there is a much bigger issue to be addressed, which is how, having catalogued all the different genomic activities in different nerve cells, the multiple types of such cells in the brain and their connections to each other, combined with an awareness of how the different regions of the brain interact at a more global level, we can thereby translate this into an awareness of how this all comes together to produce a self-conscious, thinking human brain. It is here that we may still have much to learn from the Romantic approach to the natural world, particularly the belief that the whole is present in every part of a biological system and each part is connected to the whole. One person who particularly espoused this approach in the context of human consciousness was Lev Vygotsky, whose view of the mind we discussed in Chapter 12. Vygotsky believed that science would only be able to truly decipher the material basis of consciousness by identifying a 'unit of analysis' that would be capable of reflecting all the complex interfunctional relationships that unite to produce it, from the individual nerve cell to the brain as a whole. In Vygotsky's view, such a unit of analysis had been identified in biology as a whole in the principle of natural selection, with its effects evident all the way from the conservation of a single protein sequence, through to the evolution of a whole species. However, Vygotsky argued that, for psychology, such a unit of analysis remained to be identified, as shown by its many competing and mutually exclusive explanations for how the mind worked. Today, one might add that although enormous strides have been made in the experimental neurosciences, the gap between them and the psychology of the mind remains huge.

So what might constitute such a unit of analysis for human consciousness? Based on the two central attributes that he believed constituted human uniqueness, Vygotsky believed that it must encompass both our ability to transform the world through tools and our capacity for language. But how would these two attributes manifest themselves at the cellular and genomic level in the brain? Not so long ago such a question would have been seen as a strictly one-way affair, with our genomes encoding such uniquely human attributes in the DNA sequence and transmitting this through proteins to nerve cell function. However, now that we are learning that our life experiences may significantly impact on our genomic activity, it is clear that we also need to incorporate this insight into our understanding of how the brain works. And somehow, while identifying the features of our brains that are uniquely human, we will also need to relate this to the underlying molecular and cellular mechanisms that we share with other species.

That a movement towards a more holistic way of looking at the brain is already happening should be clear from some of the cutting-edge studies that we have discussed in this book. And it is not only in this area of biology that a reconsideration of past ideas is under way. Indeed, it is surely not a coincidence that the biggest shifts in viewpoint are taking place in areas where the object of study is proving so complex that the limitations of previous approaches are becoming particularly apparent. Take, for instance, the study of metabolism, and particularly the role of organs such as the stomach, pancreas, liver, and fat tissue in this central process. Previously, the view of these different organs was of passive players controlled by signals originating in 'higher' centres, such as the hypothalamus in the brain. However, there is now an increasing recognition that each organ is an active player, sending out their own chemical signals both to each other, and back to the brain. That such a reassessment is now taking place is important given the current obesity 'epidemic', to stem which will require a more sophisticated view of the biological basis of appetite and satiety, as well as tackling the obvious social reasons for this epidemic—the surfeit of cheap, energy-rich foods, the high costs of healthier foodstuffs, and the lack of exercise that is becoming characteristic of so many in the developed world.

However, while old ways of thinking in biology are being challenged in these different areas, it would be a mistake to believe that everyone is convinced of the need for a new approach. Instead, despite emerging evidence of the intricacy of the links between our genomes and complex human diseases and characteristics,

I still attend far too many research seminars where speakers uncritically present their studies of mouse knockout 'models' of schizophrenia or autism. Apart from the difficulties of modelling these disorders in animals, given their social as well as biological component, the possibility that such single-gene knockouts might only be a crude approximation of the true genetic complexity of the human disorders, is rarely remarked upon. One reason for this may be that a complex state of affairs with multiple components is much harder to understand than one with just one, or a few, simple strands. However, it's surely a necessary step if biomedical research is to realize the promises made at the completion of the genome project and provide us with new and better means of treating disease.

Another important issue, and this, to some extent, depends upon who is correct about the extent of functionality in the human genome, is the question of how valid animal models are for understanding the true complexity of the human condition. As we've seen, one of the most surprising aspects of the ENCODE findings was the discovery that a significant amount of the biochemical activity in the genome appeared to be specific to humans. If this activity does turn out to reflect real function and not just 'noise', it would suggest that the biological differences between ourselves and other species may be greater than supposed. In fact, at the level of basic physiology there is clearly a huge amount of similarity between ourselves and a mouse, and, as such, experimentation on animals will continue to play a central role as models of human health and disease, including those of the mind. However, if we are to gain insights into the complexities of human disorders, and, as importantly, the normal human state, we will need to consider how we can supplement such studies with investigations in animals that are closer to our own species, such as other primates, but also studies in human beings themselves. In the former respect, the fact that new gene editing technologies mean it is becoming possible, for the first time, to generate knockout and other genetically modified versions of primates, offers the prospect of being able to study the effect of changes in the genome in such species. However, such a route will be controversial for some people and it will require careful consideration of the benefits of such an investigation compared to any suffering that may result in the animals being studied. Ethical considerations will be even more central to studies of human beings, whether this involves molecular analysis of the brains of dead individuals, or imaging studies of this organ in living subjects. However, one of the exciting aspects of the new genomic technologies is how much information can now be gathered about the biochemical activity in the genomes of single human cells, while safe, non-invasive imaging methods can record electrical and chemical changes across the living human brain. Particularly as we learn more about the complex interacting role of biology and environment in shaping our lives, such complements to animal studies will become increasingly important. But so will a better appreciation of the complexities of human behaviour and society, both in terms of our lives today, as well as the events that have shaped them in the past. For, ultimately, it is our self-conscious awareness and ability to shape the world around us that is the most distinctive aspect of our species.

A concern shared by many is that we seem to have lost our way with regard to this particular ability in recent years. So, while we can now sequence a human baby's genome in twenty hours or image its brain patterns as it learns to speak, at the same time other children are dying for lack of clean water or being blown up by smart missiles in some distant war. Meanwhile, despite our vast technological capacities our governments and political leaders seem incapable of doing anything to tackle the most urgent issue of our times—the warming of the planet. So perhaps it's not surprising that some people feel more threatened than empowered by the growing prospect of scientists soon being able to decode the genomes of every individual on the planet. Personally, I am excited by this prospect, but it will need to be coupled with a real and democratic debate about the ways we intend to use this information. As such, my conclusion to this book is the hope that it has stimulated a desire to know more about the workings of the genome and how it affects our lives, but also that it has reaffirmed the importance of what truly distinguishes our species—that is, our potential to not only shape the world, but to do so for the good of every human being on the planet and all the other organisms (who are, after all, our biological cousins) with whom we share this miraculous blue biosphere within the cosmos.

GLOSSARY

Chromatin Complex of DNA, histones, and non-histone proteins from which eukaryotic chromosomes are formed. Condensation during mitosis yields the visible chromosomes.

Enhancer A regulatory sequence in eukaryotic DNA that may be located far from the gene it controls. Binding of transcription factors activates transcription of the associated gene.

Enzyme A biological molecule that acts as a catalyst. Most enzymes are proteins, but certain RNAs, called ribozymes, also have catalytic activity.

Epigenetic The study of heritable changes not caused by changes in the DNA sequence; also stable alterations in the transcriptional potential of a cell that are not necessarily heritable.

Gene editing A type of genetic engineering in which DNA is inserted, replaced, or removed from a genome using artificially engineered nucleases, or 'molecular scissors'.

Gene expression Overall process by which the information encoded in a gene is converted into an observable phenotype (most commonly production of a protein).

Genome-wide association study (GWAS) Investigation of many common genetic variants in different individuals to see if any variant is associated with a characteristic.

Histones A family of small proteins found in the chromatin of all eukaryotic cells, which associate with DNA in the nucleosome.

Insulator A DNA sequence that prevents a gene regulatory protein, bound to DNA in the control region of one gene, from influencing the transcription of adjacent genes.

Meiosis In eukaryotes, a special type of cell division that occurs during maturation of the eggs and sperm.

miRNA A microRNA is a small non-coding RNA molecule found in plants, animals, and some viruses, which functions in RNA silencing and regulation of gene expression.

Mitosis In eukaryotic cells, the process whereby the nucleus is divided to produce two genetically equivalent daughter nuclei.

mRNA The messenger RNA molecule specifies the amino acid sequence of a protein. It is translated into protein in a process catalysed by ribosomes.

Mutation A permanent, heritable change in the DNA sequence of a chromosome, usually in a single gene; commonly leads to a change in or loss of the normal function of the gene product.

Nucleosome Beadlike structure in eukaryotic chromatin. Composed of a short length of DNA wrapped around a core of histones; the fundamental structural unit of chromatin.

piRNAs Non-coding RNAs linked to gene silencing of transposons in germ cells, particularly those involved in sperm formation. Recently, also identified in the brain.

Promoter DNA sequence that determines the site of transcription initiation for RNA polymerase.

Protein kinase Enzyme that transfers the phosphate group of ATP to a target protein.

Pseudogene Gene that has accumulated multiple mutations that has rendered it inactive and non-functional.

Repetitive DNA Sequences of DNA in the genome that are found to be repeated, sometimes thousands of times over.

RNA interference The phenomenon of gene silencing mediated by the interaction of a double-stranded RNA, with a corresponding target messenger RNA.

Second messenger An intracellular signalling molecule whose concentration increases (or decreases) in response to the binding of an extracellular signal to a cell-surface receptor.

Silencer A regulatory sequence in eukaryotic DNA that may be located far from the gene it controls. Binding of transcription factors inhibits transcription of the associated gene.

siRNA Small interfering RNA. Functions by causing mRNA to be inactivated after transcription, resulting in no translation.

Splicing The process by which introns are excised and exons are joined back together in the post-transcriptional modification of RNA.

Transcription Process whereby one strand of a DNA molecule is used as a template for synthesis of a complementary RNA by RNA polymerase.

Transcription factor General term for any protein, other than RNA polymerase, required to initiate or regulate transcription in eukaryotic cells.

Translation The ribosome-mediated production of a protein whose amino acid sequence is specified by the nucleotide sequence in an mRNA.

Transposition The movement of a mobile DNA element into or out of a chromosome.

ENDNOTES

Introduction: How the Genome Lost Its Junk

1. Chandrasekhar, I., Wardrop, M., and Trotman, A., Phone hacking: timeline of the scandal, *The Telegraph*, <http://www.telegraph.co.uk/news/uknews/phone-hacking/8634176/Phone-hacking-timeline-of-a-scandal.html#June12> (2012).

2. Media Fellowships, *British Science Association*, <http://www.britishscienceassociation.org/science-society/media-fellowships> (2014).

3. *Eureka Alert*, <http://www.eurekalert.org/> (2014).

4. McCrimmon, O., ENCODE data describes function of human genome, *National Human Genome Research Institute*, <http://www.genome.gov/27549810> (2012).

5. Yong, E., ENCODE: the rough guide to the human genome, *Discover Magazine*, <http://blogs.discovermagazine.com/notrocketscience/2012/09/05/encode-the-rough-guide-to-the-human-genome/#.VAS6QsVdXh4> (2012).

6. Pennisi, E., Genomics. ENCODE project writes eulogy for junk DNA. *Science* 337: 1159–61 (2012), p. 1159.

7. Whipple, T. and Parrington, J., Rummage through 'junk' DNA finds vital material, *The Times*, <http://www.thetimes.co.uk/tto/science/genetics/article3529618.ece> (2012).

8. Connor, S., Scientists debunk 'junk DNA' theory to reveal vast majority of human genes perform a vital function, *The Independent*, <http://www.independent.co.uk/news/science/scientists-debunk-junk-dna-theory-to-reveal-vast-majority-of-human-genes-perform-a-vital-function-8106777.html> (2012).

9. Hanlon, M., 'Junk DNA' and the mystery of mankind's missing genes, *The Telegraph*, <http://www.telegraph.co.uk/science/9534185/Junk-DNA-and-the-mystery-of-mankinds-missing-genes.html> (2012).

10. Jha, A., Breakthrough study overturns theory of 'junk DNA' in genome, *The Guardian*, <http://www.theguardian.com/science/2012/sep/05/genes-genome-junk-dna-encode> (2012).

11. Pennisi, E., Genomics. ENCODE project writes eulogy for junk DNA. *Science* 337: 1159–61 (2012), p. 1159.

12. John Parrington articles—media fellowships 2012, *British Science Association*, <http://www.britishscienceassociation.org/john-parrington-articles-media-fellowships-2012> (2012).

13. Graur, D., Zheng, Y., Price, N., et al., On the immortality of television sets: 'function' in the human genome according to the evolution-free gospel of ENCODE. *Genome Biology and Evolution* 5: 578–90 (2013).

14. McKie, R., Scientists attacked over claim that 'junk DNA' is vital to life, *The Guardian*, <http://www.theguardian.com/science/2013/feb/24/scientists-attacked-over-junk-dna-claim> (2013).

15. Maher, B., Fighting about ENCODE and junk, *Nature News*. <http://blogs.nature.com/news/2012/09/fighting-about-encode-and-junk.html> (2012).

16. Keller, E. F., From gene action to reactive genomes. *The Journal of Physiology* 592: 2423–9 (2014), p. 2425.

17. Morris, D., *The Naked Ape: A Zoologist's Study of the Human Animal* (Bantam Books, 1967).

18. Morris, D., *The Naked Ape: A Zoologist's Study of the Human Animal* (Bantam Books, 1967), p. 56.

19. 1967: The Naked Ape steps out, *BBC*, http://news.bbc.co.uk/onthisday/hi/dates/stories/october/12/newsid_3116000/3116329.stm (2014).

20. Pagel, M., *Wired for Culture* (Penguin, 2012), p. 81.

21. Pagel, M., *Wired for Culture* (Penguin, 2012), p. 82.

22. Barkham, P., Iraq war 10 years on: mass protest that defined a generation, *The Guardian*, <http://www.theguardian.com/world/2013/feb/15/iraq-war-mass-protest> (2013).

23. Hamer, D. H., Hu, S., Magnuson, V. L., et al., A linkage between DNA markers on the X chromosome and male sexual orientation, *Science* 261: 321–7 (1993).

24. Kitzinger, J., Constructing and deconstructing the 'gay gene': media reporting of genetics, sexual diversity and 'deviance', in *Diversity without Deviance: Human Biology, Science and Society*, ed. G. Goodman and A. Ellison, 100–17 (Taylor & Francis, 2005).

25. Mustanski, B. S., Chivers, M. L., and Bailey, J. M., A critical review of recent biological research on human sexual orientation, *Annual Review of Sex Research* 13: 89–140 (2002).

26. Radford, T., The Selfish Gene by Richard Dawkins—book review, *The Guardian*, <http://www.theguardian.com/science/2012/aug/31/the-selfish-gene-richard-dawkins-review> (2012).

27. What they said: genome in quotes, *BBC*, <http://news.bbc.co.uk/1/hi/sci/tech/807126.stm> (2000).

28. Visscher, P. M., Brown, M. A., McCarthy, M. I., and Yang, J., Five years of GWAS discovery. *American Journal of Human Genetics* 90: 7–24 (2012).

29. Li, C., Personalized medicine—the promised land: are we there yet? *Clinical Genetics* 79: 403–12 (2011).

30. Rose, H. and Rose, S., How genes failed: Hilary Rose and Steven Rose on the limitations of biological determinism, *Socialist Worker*, <http://socialistworker.co.uk/art/29639/How+genes+failed%3A+Hilary+Rose+and+Steven+Rose+on+the+limitations+of+biological+determinism> (2012).

31. Van Regenmortel, M. H. V., Reductionism and complexity in molecular biology. *EMBO Reports* 5: 1016–20 (2004), p. 1016.

32. Parrington, J. and Coward, K., The spark of life. *The Biologist* 50: 5–10 (2003).

33. Parrington, J., Davis, L. C., Galione, A., and Wessel, G., Flipping the switch: how a sperm activates the egg at fertilization. *Developmental Dynamics* 236: 2027–38 (2007).

34. Doyle, A., McGarry, M. P., Lee, N. A., and Lee, J. J., The construction of transgenic and gene knockout/knockin mouse models of human disease. *Transgenic Research* 21: 327–49 (2012).

35. Fryer, R. M., Randall, J., Yoshida, T., et al., Global analysis of gene expression: methods, interpretation, and pitfalls. *Experimental Nephrology* 10: 64–74 (2002).

36. Levine, M., Cattoglio, C., and Tjian, R., Looping back to leap forward: transcription enters a new era. *Cell* 157: 13–25 (2014).

37. Gibcus, J. H. and Dekker, J. The hierarchy of the 3D genome. *Molecular Cell* 49: 773–82 (2013).

38. Morris, K. V. and Mattick, J. S., The rise of regulatory RNA. *Nature Reviews Genetics* 15: 423–37 (2014).

39. Rivera, C. M. and Ren, B., Mapping human epigenomes. *Cell* 155: 39–55 (2013).

40. Slotkin, R. K. and Martienssen, R., Transposable elements and the epigenetic regulation of the genome. *Nature Reviews Genetics* 8: 272–85 (2007).

41. Lupski, J. R., Genetics: genome mosaicism—one human, multiple genomes. *Science* 341: 358–9 (2013).

42. Reilly, M. T., Faulkner, G. J., Dubnau, J., et al., The role of transposable elements in health and diseases of the central nervous system. *The Journal of Neuroscience* 33: 17577–86 (2013).

43. Stindl, R., The telomeric sync model of speciation: species-wide telomere erosion triggers cycles of transposon-mediated genomic rearrangements, which underlie the saltatory appearance of nonadaptive characters. *Die Naturwissenschaften* 101: 163–86 (2014).

44. Pääbo, S., *Neanderthal Man* (Basic Books, 2014).

45. Ahmed, M. and Liang, P., Study of modern human evolution via comparative analysis with the Neanderthal genome. *Genomics & Informatics* 11: 230–8 (2013).

46. Li, C., Personalized medicine—the promised land: are we there yet? *Clinical Genetics* 79: 403–12 (2011).

Chapter 1: The Inheritors

47. Comparable odds to winning a Lotto jackpot, *Lottery*, <http://lottery.typepad.com/lottery/2012/06/comparable-odds-to-winning-a-lotto-jackpot.html> (2012).

48. Spector, D., The odds of you being alive are incredibly small, *Business Insider*, <http://www.businessinsider.com/infographic-the-odds-of-being-alive-2012-6> (2012).

49. Darwin, C., *Origin of Species* (John Murray, 1859).

50. Darwin, C., *Origin of Species* (John Murray, 1859), p. 78.

51. Shermer, M., *In Darwin's Shadow* (Oxford University Press, 2002), p. 161.

52. Young, R. M., Malthus on man—in animals no moral restraint. *Clio Medica* 59: 73–91 (2000).

53. Desmond, A. and Moore, J. R., *Darwin* (Michael Joseph, 1991), p. 468.

54. Desmond, A. and Moore, J. R., *Darwin* (Michael Joseph, 1991), p. 470.

55. Desmond, A. and Moore, J. R., *Darwin* (Michael Joseph, 1991), p. 469.

56. Darwin, C., *The Descent of Man* (John Murray, 1871), p. 485.

57. Shermer, M., *In Darwin's Shadow* (Oxford University Press, 2002), p. 159.

58. Gould, S. J. and Berry, A., *Infinite Tropics: An Alfred Russel Wallace Anthology* (Verso, 2003), p. 208.

59. Shermer, M., *In Darwin's Shadow* (Oxford University Press, 2002), p. 161.

60. Lennox, J., Aristotle's Biology, *Stanford Encylopedia of Philosophy*, <http://plato.stanford.edu/entries/aristotle-biology/> (2014).

61. Gregor Johann Mendel, *Complete Dictionary of Scientific Biography*, <http://www.encyclopedia.com/topic/Gregor_Johann_Mendel.aspx> (2008).

62. Badano, J. L. and Katsanis, N., Beyond Mendel: an evolving view of human genetic disease transmission. *Nature Reviews Genetics* 3: 779–89 (2002).

63. Watson, J. D., *DNA* (Arrow Books, 2004), p. 10.

64. Lenay, C., Hugo de Vries: from the theory of intracelullar pangenesis to the rediscovery of Mendel. *Life Sciences* 323: 1053–60 (2000).

65. Wallace, A., *Letters and Reminiscences* (Cassell, 1916), p. 108.

66. Kohler, R. E., *Lords of the Fly: Drosophila Genetics and the Experimental Life* (University of Chicago Press, 1994), p. 41.

67. Bowler, P. J., Hugo De Vries and Thomas Hunt Morgan: the mutation theory and the spirit of Darwinism. *Annals of Science* 35: 55–73 (1978).

68. Genes, Chromosomes, and the Origins of Modern Biology, *Columbia University*, <http://www.columbia.edu/cu/alumni/Magazine/Legacies/Morgan/> (2014).

69. Watson, J. D., *DNA* (Arrow Books, 2004), p. 14.

70. Paweletz, N., Walther Flemming: pioneer of mitosis research. *Nature Reviews Molecular and Cellular Biology* 2: 72–5 (2001).

71. Laubichler, M. D. and Davidson, E. H., Boveri's long experiment: sea urchin merogones and the establishment of the role of nuclear chromosomes in development. *Developmental Biology* 314: 1–11 (2008).

72. Wessel, G., Y does it work this way? Nettie Maria Stevens (July 7, 1861–May 4, 1912). *Molecular Reproduction and Development* 78 (2011).

73. Schramm, W., The history of haemophilia—a short review. *Thrombosis Research* 134S1: S4-S9 (2014).

74. Benson, K. R., T. H. Morgan's resistance to the chromosome theory. *Nature Reviews Genetics* 2: 469–74 (2001).

75. Carlson, E. A., H. J. Muller's contributions to mutation research. *Mutation Research* 752: 1–5 (2013).

76. Portin, P., The birth and development of the DNA theory of inheritance: sixty years since the discovery of the structure of DNA. *Journal of Genetics* 93: 293–302 (2014).

77. Koszul, R., Meselson, M., Van Doninck, K., et al., The centenary of Janssens's chiasmatype theory. *Genetics* 191: 309–17 (2012).

78. Brush, S. G., How theories became knowledge: Morgan's chromosome theory of heredity in America and Britain. *Journal of the History of Biology* 35: 471–535 (2002).

79. Maas, W., *Gene Action: A Historical Account* (Oxford University Press, 2002), p. 16.

80. Scriver, C. R., Garrod's foresight; our hindsight. *Journal of Inherited Metabolic Disease* 24: 93–116 (2001).

81. Edwards, A. W., Mathematizing Darwin. *Behavioral Ecology and Sociobiology* 65: 421–30 (2011).

82. Rao, V. and Nanjundiah, V., J. B. S. Haldane, Ernst Mayr and the Beanbag genetics dispute. *Journal of the History of Biology* 44: 233–81 (2011).

83. Haldane, J. B., A defense of beanbag genetics. 1964. *International Journal of Epidemiology* 37: 435–42 (2008).

84. Singer, M. and Berg, P., George Beadle: from genes to proteins. *Nature Reviews in Genetics* 5: 949–54 (2004).

Chapter 2: Life as a Code

85. Kean, S., *The Violinist's Thumb* (Doubleday, 2012), p. 18.

86. Dahm, R., Friedrich Miescher and the discovery of DNA. *Developmental Biology* 278: 274–88 (2005), p. 279.

87. Judson, H. F., *The Eighth Day of Creation* (Simon & Schuster, 1979), p. 30.

88. Watson, J. D., *DNA* (Arrow Books, 2004), p. 38.

89. Judson, H. F., *The Eighth Day of Creation* (Simon & Schuster, 1979), p. 59.

90. The Martha Chase Effect: Part 1, *Sciopic*, <http://sciopic.wordpress.com/2013/05/16/the-martha-chase-effect-part-1/> (2013).

91. *The Literature Network*, <http://www.online-literature.com/swift/3515/> (2014).

92. Watson, J. D., *The Annotated and Illustrated Double Helix* (Simon & Schuster, 2012).

93. Watson, J. D., *Avoid Boring People and Other Lessons from a Life in Science* (Oxford University Press, 2007), p. 39.

94. *James Watson: The double helix and beyond*, <http://www.npr.org/2012/11/16/165278526/james-watson-the-double-helix-and-beyond> (2012).

95. Landau, E., Watson: 'DNA was my only gold rush', *CNN*, <http://edition.cnn.com/2013/06/28/health/james-watson-dna/> (2013).

96. Watson, J. D., *The Annotated and Illustrated Double Helix* (Simon & Schuster, 2012), p. 37.

97. Jenkin, J., A unique partnership: William and Lawrence Bragg and the 1915 Nobel Prize in Physics. *Minerva* 39: 373–92 (2001).

98. Maddox, B., *Rosalind Franklin: The Dark Lady of DNA* (Harper Collins, 2002), p. 125.

99. The Randall letters: the DNA story at King's revisited, *King's College London Archive*, <http://dnaandsocialresponsibility.blogspot.co.uk/2010/08/randall-letters-dna-story-at-kings.html> (2010).

100. Watson, J. D., *The Annotated and Illustrated Double Helix* (Simon & Schuster, 2012), p. 182.

101. Wilkins, M., *The Third Man of the Double Helix* (Oxford University Press, 2003).

102. Maddox, B., *Rosalind Franklin: The Dark Lady of DNA* (Harper Collins, 2002), p. 127.

103. Watson, J. D., *The Annotated and Illustrated Double Helix* (Simon & Schuster, 2012), p. 200.

104. Watson, J. D., *The Annotated and Illustrated Double Helix* (Simon & Schuster, 2012), p. 209.

105. Kemp, M., The Mona Lisa of modern science. *Nature* 421: 416–20 (2003), p. 416.

106. Portin, P., The birth and development of the DNA theory of inheritance: sixty years since the discovery of the structure of DNA. *Journal of Genetics* 93: 293–302 (2014).

107. Judson, H. F., *The Eighth Day of Creation* (Simon & Schuster, 1979), p. 142.

108. Rose, S., Practicing biochemistry without a licence? *EMBO Reports* 12: 381 (2011), p. 381.

109. Watson, J. D. and Crick, F. H. C., Molecular structure of nucleic acids: a structure for deoxyribose nucleic acid. *Nature* 171: 737–8 (1953), p. 737.

110. Watson, J. D. and Crick, F. H. C., Genetical implications of the structure of deoxyribonucleic acid. *Nature* 171: 964–7 (1953).

111. Linus Pauling and the race for DNA, *Oregon State University Libraries*, <http://scarc.library.oregonstate.edu/coll/pauling/dna/quotes/all.html> (2014).

112. Maddox, B., *Rosalind Franklin: The Dark Lady of DNA* (Harper Collins, 2002), p. 307.

113. Sydney Brenner—biographical, *Nobel Media*, <http://www.nobelprize.org/nobel_prizes/medicine/laureates/2002/brenner-bio.html> (2002).

114. Watson, J. D., *DNA* (Arrow Books, 2004), p. 56.

115. Davis, T. H., Meselson and Stahl: the art of DNA replication. *Proceedings of the National Academy of Sciences of the United States of America* 101: 17895–6 (2004).

116. Lehman, I. R., Historical perspective: Arthur Kornberg, a giant of twentieth-century biochemistry. *Trends in Biochemical Sciences* 33: 291–6 (2008).

117. Stretton, A. O. W., The first sequence: Fred Sanger and insulin. *Genetics* 162: 527–32 (2002).

118. Judson, H., *The Eighth Day of Creation* (Simon & Schuster, 1979), p. 213.

119. Segre, G., The Big Bang and the genetic code. *Nature* 404: 437 (2000).

120. Gamow, G., *The Creation of the Universe* (Courier Dover Publications, 2004), p. 139.

121. Watson, J. D., *DNA* (Arrow Books, 2004), p. 64.

122. Watson, J. D., *DNA* (Arrow Books, 2004), p. 66.

123. Kresge, N., Simoni, R. D., and Hill, R., The discovery and isolation of RNA polymerase by Jerard Hurwitz. *Journal of Biological Chemistry* 281: e12–e14 (2006).

124. Watson, J. D., *DNA* (Arrow Books, 2004), p. 67.

125. Judson, H. F., *The Eighth Day of Creation* (Simon & Schuster, 1979), p. 282.

126. Pederson, T., Obituary: Paul C. Zamecnik (1912–2009). *Nature* 462: 423 (2009).

127. Sydney Brenner—biographical, *Nobel Media*, <http://www.nobelprize.org/nobel_prizes/medicine/laureates/2002/brenner-bio.html> (2002).

128. Pieribone, V., *Aglow in the Dark: The Revolutionary Science of Biofluorescence* (Harvard University Press, 2006), p. 113.

129. Judson, H. F., *The Eighth Day of Creation* (Simon & Schuster, 1979), pp. 436–41.

130. Shaw, K., The role of ribosomes in protein synthesis. *Nature Education* 1: 201 (2008).

131. Yanofsky, C., Establishing the triplet nature of the genetic code. *Cell* 128: 815–18 (2007).

132. Nazarali, A. J., Marshall Nirenberg 1927–2010. *Cellular and Molecular Neurobiology* 31: 805–7 (2011).

133. Crick, F. H. C., *What Mad Pursuit* (Basic Books, 1990), p. 109.

134. Judson, H. F., *The Eighth Day of Creation* (Simon & Schuster, 1979), p. 41.

135. Lewontin, R. C., *The Triple Helix* (Harvard University Press, 2000), p. 10.

136. Collins, N., Sir John Gurdon, Nobel Prize winner, was 'too stupid' for science at school, *The Telegraph*, <http://www.telegraph.co.uk/science/science-news/9594351/Sir-John-Gurdon-Nobel-Prize-winner-was-too-stupid-for-science-at-school.html> (2012).

137. Gurdon, J., Nuclear reprogramming in eggs. *Nature Medicine* 15: 1141–4 (2009).

138. Stanford, P. K., August Weismann's theory of the germ-plasm and the problem of unconceived alternatives. *History and Philosophy of the Life Sciences* 27: 163 (2005).

139. Gurdon, J. B., Interview with Sir John B. Gurdon, *Nobel Media*, <http://www.nobelprize.org/nobel_prizes/medicine/laureates/2012/gurdon-telephone.html> (2012).

Chapter 3: Switches and Signals

140. Jayaraman, R., Jacques Monod and the advent of the age of operons. *Resonance* 15: 1084–96 (2010).

141. Judson, H. F., *The Eighth Day of Creation* (Simon & Schuster, 1979), p. 357.

142. Jayaraman, R., Jacques Monod and the advent of the age of operons. *Resonance* 15: 1084–96 (2010), p. 1085.

143. Carroll, S. B., *Brave Genius: A Scientist, a Philosopher, and their Daring Adventures from the French Resistance to the Nobel Prize* (Broadway Books, 2014).

144. Watson, J. D., *DNA* (Arrow Books, 2004), p. 77.

145. Ullmann, A., In memoriam: Jacques Monod (1910–1976). *Genome Biology and Evolution* 3: 1025–33 (2011).

146. François Jacob, *The Telegraph*, <http://www.telegraph.co.uk/news/obituaries/10076471/Francois-Jacob.html> (2013).

147. Judson, H. F., *The Eighth Day of Creation* (Simon & Schuster, 1979), p. 402.

148. Judson, H. F., *The Eighth Day of Creation* (Simon & Schuster, 1979), p. 416.

149. Kay, L. E., *Who Wrote the Book of Life? A History of the Genetic Code* (Stanford University Press, 2000), p. 215.

150. Judson, H. F., *The Eighth Day of Creation* (Simon & Schuster, 1979), p. 373.

151. Englesberg, E. and Wilcox, G., Regulation: positive control. *Annual Review of Genetics* 8: 219–42 (1974).

152. Bussell, K., Accentuate the positive. *Nature Milestones*, S7, <http://www.nature.com/milestones/geneexpression/milestones/articles/milegene04.html> (2005).

153. Carroll, S. B., The main characters, *Sean B. Carroll website*, <http://seanbcarroll.com/main-characters/> (2014).

154. Muller-Hill, B., *The Lac Operon: A Short History of a Genetic Paradigm* (De Gruyter 1996).

155. Judson, H. F., *The Eighth Day of Creation* (Simon & Schuster, 1979), p. 418.
156. Ullmann, A., In memoriam: Jacques Monod (1910–1976). *Genome Biology and Evolution* 3: 1025–33 (2011), p. 1029.
157. Kresge, N., Simoni, R. D., and Hill, R., Earl W. Sutherland's discovery of cyclic adenine monophosphate and the second messenger system. *Journal of Biological Chemistry* 280: e39–40 (2005).
158. Hicks, J., Wartchow, E., and Mierau, G., Glycogen storage diseases: a brief review and update on clinical features, genetic abnormalities, pathologic features, and treatment. *Ultrastructural Pathology* 35: 183–96 (2011).
159. Rajfer, J., Discovery of NO in the penis. *International Journal of Impotence Research* 20: 431–6 (2008).
160. Rajfer, J., Discovery of NO in the penis. *International Journal of Impotence Research* 20: 431–6 (2008), p. 432.
161. Newman, R. H., Zhang, J., and Zhu, H., Toward a systems-level view of dynamic phosphorylation networks. *Frontiers in Genetics* 5: 263 (2014).
162. Sands, W. A. and Palmer, T. M., Regulating gene transcription in response to cyclic AMP elevation. *Cellular Signalling* 20: 460–6 (2008).
163. Villeda, S. A., Plambeck, K. E., Middeldorp, J., et al., Young blood reverses age-related impairments in cognitive function and synaptic plasticity in mice. *Nature Medicine* 20: 659–63 (2014).
164. Sample, I., Infusions of young blood may reverse effects of ageing, studies suggest, *The Guardian*, <http://www.theguardian.com/science/2014/may/04/young-blood-reverse-ageing-mice-studies> (2014).
165. Beckwith, J., Fifty years fused to lac. *Annual Review of Microbiology* 67: 1–19 (2013).
166. Kevles, D., Biologist at the barricades, *American Scientist*, <http://www.americanscientist.org/bookshelf/pub/biologist-at-the-barricades> (2003).
167. How the past informs the future, *Bright Boys*, <http://www.brightboys.org/PDF/Innovation_on_the_Verge.pdf> (2014).
168. Judson, H. F., *The Eighth Day of Creation* (Simon & Schuster, 1979), p. 588.
169. Judson, H. F., *The Eighth Day of Creation* (Simon & Schuster, 1979), p. 585.
170. Judson, H. F., *The Eighth Day of Creation* (Simon & Schuster, 1979), p. 590.
171. Watson, J. D., *Avoid Boring People and Other Lessons from a Life in Science* (Oxford University Press, 2007), pp. 257–8.
172. Burley, S. K., X-ray crystallographic studies of eukaryotic transcription initiation factors. *Philosophical Transactions of the Royal Society of London. Series B, Biological Sciences* 351: 483–9 (1996).
173. Gurdon, J. B. and Melton, D. A., Nuclear reprogramming in cells. *Science* 322: 1811–15 (2008).
174. Dahm, R., Friedrich Miescher and the discovery of DNA. *Developmental Biology* 278: 274–88 (2005).
175. Olins, D. E. and Olins, A. L., Chromatin history: our view from the bridge. *Nature Reviews Molecular Cell Biology* 4: 809–14 (2003).
176. Maddox, B., *Rosalind Franklin: The Dark Lady of DNA* (Harper, 2003), p. 229.
177. Travers, A. A. and Klug, A., The bending of DNA in nucleosomes and its wider implications. *Philosophical Transactions of the Royal Society B: Biological Sciences* 317: 537–61 (1987).
178. Klug, A., Aaron Klug's speech at the Nobel banquet, December 10, 1982, *Nobel Media*, <http://www.nobelprize.org/nobel_prizes/chemistry/laureates/1982/klug-speech.html> (1982).

179. Shahbazian, M. D. and Grunstein, M., Functions of site-specific histone acetylation and deacetylation. *Annual Review of Biochemistry* 76: 75–100 (2007).

180. Zimmer, C., E. coli: why I am in love with a bacterium, *The Telegraph*, <http://www. telegraph.co.uk/science/science-news/3345320/E.-coli-why-I-am-in-love-with-a-bacterium. html> (2008).

Chapter 4: The Spacious Genome

181. Hibbing, M. E., Fuqua, C., Parsek, M. R., and Peterson, S. B., Bacterial competition: surviving and thriving in the microbial jungle. *Nature Reviews Microbiology* 8: 15–25 (2010).

182. Lever, M. A., Rouxel, O., Alt, J. C., et al., Evidence for microbial carbon and sulfur cycling in deeply buried ridge flank basalt. *Science* 339: 1305–8 (2013).

183. Jayaraman, R., Jacques Monod and the advent of the age of operons. *Resonance* 15: 1084–96 (2010).

184. Weiss, R. A. and Vogt, P. K., 100 years of Rous sarcoma virus. *The Journal of Experimental Medicine* 208: 2351–5 (2011).

185. Mammas, I. N., Sourvinos, G., and Spandidos, D. A., The paediatric story of human papillomavirus. *Oncology Letters* 8: 502–6 (2014).

186. Khoury, G. and Gruss, P., Enhancer elements. *Cell* 33: 313–14 (1983).

187. Shlyueva, D., Stampfel, G., and Stark, A., Transcriptional enhancers: from properties to genome-wide predictions. *Nature Reviews Genetics* 15: 272–86 (2014).

188. Nowotschin, S. and Hadjantonakis, A. K., Cellular dynamics in the early mouse embryo: from axis formation to gastrulation. *Current Opinion in Genetics & Development* 20: 420–7 (2010).

189. Blum, M., Feistel, K., Thumberger, T., and Schweickert, A., The evolution and conservation of left-right patterning mechanisms. *Development* 141: 1603–13 (2014).

190. Notable persons with situs inversus, *My Hallucination*, <http://srsekhar.blogspot.co.uk/2008/10/notable-persons-with-situs-inversus.html> (2008).

191. Unique anatomy: situs inversus and interrupted IVC, *Into Another World*, <http://susanleighnoble.wordpress.com/2013/05/20/unique-anatomy-situs-inversus-interrupted-ivc/> (2013).

192. One-year-old Indian boy breaks world record after being born with THIRTY FOUR fingers and toes, *The Daily Mail*, <http://www.dailymail.co.uk/news/article-2018470/Akshat-Saxena-Indian-boy-born-34-fingers-toes-breaks-world-record.html> (2011).

193. Mallo, M., Wellik, D. M., and Deschamps, J., Hox genes and regional patterning of the vertebrate body plan. *Developmental Biology* 344: 7–15 (2010).

194. Smith, E. and Shilatifard, A., Enhancer biology and enhanceropathies. *Nature Structural & Molecular Biology* 21: 210–19 (2014).

195. Levine, M., Cattoglio, C., and Tjian, R., Looping back to leap forward: transcription enters a new era. *Cell* 157: 13–25 (2014).

196. Gregory, T. R., Coincidence, coevolution, or causation? DNA content, cell size, and the C-value enigma. *Biological Reviews* 76: 65–101 (2001).

197. Zimmer, C., The Case for Junk DNA, *National Geographic*, <http://phenomena.nationalgeographic.com/2014/05/09/the-case-for-junk-dna/> (2014).

198. Sharp, P. A., The discovery of split genes and RNA splicing. *Trends in Biochemical Sciences* 30: 279–81 (2005).

199. Breathnach, R. and Chambon, P., Organization and expression of eukaryotic split genes coding for proteins. *Annual Review of Biochemistry* 50: 349–83 (1981).

200. Hertel, K. J., Combinatorial control of exon recognition. *The Journal of Biological Chemistry* 283: 1211–15 (2008).

201. Gamazon, E. R. and Stranger, B. E., Genomics of alternative splicing: evolution, development and pathophysiology. *Human Genetics* 133: 679–87 (2014).

202. Halaby, D. M., Poupon, A., and Mornon, J. P., The immunoglobulin fold family: sequence analysis and 3D structure comparisons. *Protein Engineering* 12: 563–71 (1999).

203. Peterson, M. L., Mechanisms controlling production of membrane and secreted immunoglobulin during B cell development. *Immunologic Research* 37: 33–46 (2007).

204. Warner, B., Charles Darwin and John Herschel. *South African Journal of Science* 105: 432–3 (2009).

205. Hoyle, F., *Intelligent Universe: A New View of Creation and Evolution* (Michael Joseph Ltd., 1983).

206. Liu, M. and Grigoriev, A., Protein domains correlate strongly with exons in multiple eukaryotic genomes—evidence of exon shuffling? *Trends in Genetics: TIG* 20: 399–403 (2004).

207. Lewandowska, M. A., The missing puzzle piece: splicing mutations. *International Journal of Clinical and Experimental Pathology* 6: 2675–82 (2013).

208. Kaplan, F., Tay–Sachs disease carrier screening: a model for prevention of genetic disease. *Genetic Testing* 2: 271–92 (1998).

209. Mahuran, D. J., Biochemical consequences of mutations causing the GM2 gangliosidoses. *Biochimica et Biophysica Acta* 1455: 105–38 (1999).

210. George, A., The rabbi's dilemma, *New Scientist*, <http://www.newscientist.com/article/mg18124345.400-the-rabbis-dilemma.html?full=true#.VAg-Jo10wuQ> (2004).

211. Sanger, F., Sequences, sequences, and sequences. *Annual Reviews in Biochemistry* 57: 1–28 (1988).

212. Sulston, J. and Ferry, G., *The Common Thread* (Corgi, 2009).

213. Ohno, S., So much 'junk' DNA in our genome. *Brookhaven Symposium on Biology* 23: 366–70 (1972).

214. Dawkins, R., *The Selfish Gene: 30th Anniversary Edition* (Oxford University Press, 2006), p. 45.

215. Orgel, L. E. and Crick, F. H., Selfish DNA: the ultimate parasite. *Nature* 284: 604–7 (1980).

216. Clark, K. R., *Religion and the Sciences of Origins* (Palgrave Macmillan, 2014), p. 65.

217. Lamb, T. D., Collin, S. P., and Pugh, E. N., Jr., Evolution of the vertebrate eye: opsins, photoreceptors, retina and eye cup. *Nature Reviews Neuroscience* 8: 960–76 (2007).

218. Gould, S. J., *The Panda's Thumb: More Reflections in Natural History* (Penguin, 1990).

219. Dawkins, R., *The Greatest Show on Earth: The Evidence for Evolution* (Black Swan, 2010), p. 332.

220. Miller, K. R., Life's grand design. *Technology Review*, <http://www.millerandlevine.com/km/evol/lgd/> (1994).

Chapter 5: RNA Out of the Shadows

221. Darwin, C., *Origin of Species* (John Murray, 1859), p. 484.

222. Follmann, H. and Brownson, C., Darwin's warm little pond revisited: from molecules to the origin of life. *Die Naturwissenschaften* 96: 1265–92 (2009), p. 1265.

223. Tirard, S., Origin of life and definition of life, from Buffon to Oparin. *Origins of Life and Evolution of the Biosphere: The Journal of the International Society for the Study of the Origin of Life* 40: 215–20 (2010).

224. Miller, S. L. and Urey, H. C., Origin of life. *Science* 130: 1622–4 (1959).

225. Pääbo, S., *Neanderthal Man* (Basic Books, 2014).

226. Neveu, M., Kim, H. J., and Benner, S. A., The 'strong' RNA world hypothesis: fifty years old. *Astrobiology* 13: 391–403 (2013).

227. Staley, J. P. and Woolford, J. L., Jr., Assembly of ribosomes and spliceosomes: complex ribonucleoprotein machines. *Current Opinion in Cell Biology* 21: 109–18 (2009).

228. An introduction to enzymes, *Broad Institute*, <http://www.broadinstitute.org/~rivas/www/Biochem/enz.pdf> (2014).

229. Cech, T. R., Nobel lecture. Self-splicing and enzymatic activity of an intervening sequence RNA from Tetrahymena. *Bioscience Reports* 10: 239–61 (1990).

230. Altman, S., Nobel lecture. Enzymatic cleavage of RNA by RNA. *Bioscience Reports* 10: 317–37 (1990).

231. Korostelev, A. and Noller, H. F., The ribosome in focus: new structures bring new insights. *Trends in Biochemical Sciences* 32: 434–41 (2007).

232. Sidney Altman—biographical, *Nobel Media*, <http://www.nobelprize.org/nobel_prizes/chemistry/laureates/1989/altman-bio.html> (1989).

233. Levin, R. C., *The Work of the University* (Yale University Press, 2008), p. 42.

234. Ellington, A. D., Chen, X., Robertson, M., and Syrett, A., Evolutionary origins and directed evolution of RNA. *The International Journal of Biochemistry & Cell Biology* 41: 254–65 (2009).

235. Faculty Member—Jack Szostak, *Harvard University*, <http://dms.hms.harvard.edu/bbs/fac/Szostak.php> (2014).

236. Deamer, D. W., Origins of life: how leaky were primitive cells? *Nature* 454: 37–8 (2008).

237. Watson, J. D., *DNA* (Arrow Books, 2004), p. 81.

238. Napoli, C., Lemieux, C., and Jorgensen, R., Introduction of a chimeric chalcone synthase gene into petunia results in reversible co-suppression of homologous genes in trans. *The Plant Cell* 2: 279–89 (1990).

239. Billmyre, R. B., Calo, S., Feretzaki, M., et al., RNAi function, diversity, and loss in the fungal kingdom. *Chromosome Research: An International Journal on the Molecular, Supramolecular and Evolutionary Aspects of Chromosome Biology* 21: 561–72 (2013).

240. Sommer, R. J. and Bumbarger, D. J., Nematode model systems in evolution and development. *Wiley Interdisciplinary Reviews. Developmental Biology* 1: 389–400 (2012).

241. Putcha, G. V. and Johnson, E. M., 'Men are but worms': neuronal cell death in C. elegans and vertebrates. *Cell Death and Differentiation* 11: 38–48 (2004).

242. Shaikh, S. and Leonard-Amodeo, J., The deviating eyes of Michelangelo's David. *Journal of the Royal Society of Medicine* 98: 75–6 (2005), p. 75.

243. Fire, A., Xu, S., Montgomery, M. K., et al., Potent and specific genetic interference by double-stranded RNA in Caenorhabditis elegans. *Nature* 391: 806–11 (1998).

244. Craig C. Mello—biographical, *Nobel Media*, <http://www.nobelprize.org/nobel_prizes/medicine/laureates/2006/mello-bio.html> (2006).

245. Kurreck, J., RNA interference: from basic research to therapeutic applications. *Angewandte Chemie* 48: 1378–98 (2009).

246. Conger, K., Andrew Fire wins 2006 Nobel Prize in Physiology or Medicine, *Stanford University News Service*, <http://news.stanford.edu/pr/2006/pr-nobel-100206.html> (2006).

247. Campeau, E. and Gobeil, S., RNA interference in mammals: behind the screen. *Briefings in Functional Genomics* 10: 215–26 (2011).

248. Espiritu, M. J., Collier, A. C., and Bingham, J. P., A twenty-first-century approach to age-old problems: the ascension of biologics in clinical therapeutics. *Drug Discovery Today* 19: 1109–13 (2014).

249. Zhou, Y., Zhang, C., and Liang, W., Development of RNAi technology for targeted therapy—a track of siRNA based agents to RNAi therapeutics. *Journal of Controlled Release: Official Journal of the Controlled Release Society* 193: 270–81 (2014).

250. Incarbone, M. and Dunoyer, P., RNA silencing and its suppression: novel insights from in planta analyses. *Trends in Plant Science* 18: 382–92 (2013).

251. Sun, K. and Lai, E. C., Adult-specific functions of animal microRNAs. *Nature Reviews Genetics* 14: 535–48 (2013).

252. Watanabe, T. and Lin, H., Posttranscriptional regulation of gene expression by piwi proteins and piRNAs. *Molecular Cell* 56: 18–27 (2014).

253. Kapusta, A. and Feschotte, C., Volatile evolution of long noncoding RNA repertoires: mechanisms and biological implications. *Trends in Genetics* 30: 439–52 (2014).

254. Arora, S., Rana, R., Chhabra, A., et al., miRNA-transcription factor interactions: a combinatorial regulation of gene expression. *Molecular Genetics and Genomics* 288: 77–87 (2013).

255. Gama Sosa, M. A., De Gasperi, R., and Elder, G. A., Animal transgenesis: an overview. *Brain Structure & Function* 214: 91–109 (2010).

256. Van den Driesche, S., Sharpe, R. M., Saunders, P. T., and Mitchell, R. T., Regulation of the germ stem cell niche as the foundation for adult spermatogenesis: a role for miRNAs? *Seminars in Cell & Developmental Biology* 29: 76–83 (2014).

257. Undi, R. B., Kandi, R., and Gutti, R. K., MicroRNAs as haematopoiesis regulators. *Advances in Hematology* 2013: 695754 (2013).

258. Hesselberth, J. R., Lives that introns lead after splicing. *Wiley Interdisciplinary Reviews in RNA* 4: 677–91 (2013).

Chapter 6: It's a Jungle in There!

259. Rose, H. and Rose, S., *Genes, Cells and Brains* (Verso, 2012), p. 25.

260. Consortium, T. E. P., An integrated encyclopedia of DNA elements in the human genome. *Nature* 489: 57–74 (2012).

261. Brenner, S. Loose ends. *Current Biology* 5: 332 (1995).

262. Johnson, J. M., Edwards, S., Shoemaker, D., and Schadt, E. E., Dark matter in the genome: evidence of widespread transcription detected by microarray tiling experiments. *Trends in Genetics* 21: 93–102 (2005).

263. Ostriker, J. P. and Steinhardt, P., New light on dark matter. *Science* 300: 1909–13 (2003).

264. Sample, I., Dark matter may have been detected—streaming from the sun's core, *The Guardian*, <http://www.theguardian.com/science/2014/oct/16/dark-matter-detected-sun-axions> (2014).

265. Thomas, D. J., Rosenbloom, K. R., Clawson, H., et al., The ENCODE Project at UC Santa Cruz. *Nucleic Acids Research* 35: D663–7 (2007).

266. Rothbart, S. B. and Strahl, B. D., Interpreting the language of histone and DNA modifications. *Biochimica et Biophysica Acta* 1839: 627–43 (2014).

267. Dogini, D. B., Pascoal, V. D. B., Avansini, S. H., et al., The new world of RNAs. *Genetics and Molecular Biology* 37: 285–93 (2014).

268. Yong, E., ENCODE: the rough guide to the human genome, *Discover Magazine*, <http://blogs.discovermagazine.com/notrocketscience/2012/09/05/encode-the-rough-guide-to-the-human-genome/#.VAS6QsVdXh4> (2012).

269. Pennisi, E., Genomics. ENCODE project writes eulogy for junk DNA, *Science* 337: 1159–61 (2012), p. 1159.

270. Pennisi, E., Genomics. ENCODE project writes eulogy for junk DNA, *Science* 337: 1159–61 (2012), p. 1161.

271. Visscher, P. M., Brown, M. A., McCarthy, M. I., and Yang, J., Five years of GWAS discovery. *American Journal of Human Genetics* 90: 7–24 (2012).

272. Van der Sijde, M. R., Ng, A., and Fu, J., Systems genetics: from GWAS to disease pathways. *Biochimica et Biophysica Acta* 1842: 1903–1909 (2014).

273. Stamatoyannopoulos, J. A., What does our genome encode? *Genome Research* 22: 1602–11 (2012).

274. Whipple, T. and Parrington, J., Rummage through 'junk' DNA finds vital material, *The Times*, <http://www.thetimes.co.uk/tto/science/genetics/article3529618.ece> (2012).

275. Ward, L. D. and Kellis, M., Evidence of abundant purifying selection in humans for recently acquired regulatory functions. *Science* 337: 1675–8 (2012).

276. Graur, D., Zheng, Y., Price, N., et al., On the immortality of television sets: 'function' in the human genome according to the evolution-free gospel of ENCODE, *Genome Biology and Evolution* 5: 578–90 (2013).

277. Graur, D., Zheng, Y., Price, N., et al., On the immortality of television sets: 'function' in the human genome according to the evolution-free gospel of ENCODE, *Genome Biology and Evolution* 5: 578–90 (2013), p. 578.

278. McKie, R., Scientists attacked over claim that 'junk DNA' is vital to life, *The Guardian*, <http://www.theguardian.com/science/2013/feb/24/scientists-attacked-over-junk-dna-claim> (2013).

279. Bhattacharjee, Y., The vigilante. *Science* 343: 1306–9 (2014).

280. Spencer, G., New Genome Comparison Finds Chimps, Humans Very Similar at the DNA Level, *National Human Genome Research Institute*, <http://www.genome.gov/15515096> (2005).

281. Why mouse matters, *National Human Genome Research Institute*, <http://www.genome.gov/10001345> (2010).

282. DNA, *Natural History Museum*, <http://www.nhm.ac.uk/nature-online/evolution/what-is-the-evidence/morphology/dna-molecules/> (2014).

283. Haerty, W. and Ponting, C. P., No gene in the genome makes sense except in the light of evolution. *Annual Review of Genomics and Human Genetics* 15: 71–92 (2014).

284. Graur, D., Zheng, Y., Price, N., et al., On the immortality of television sets: 'function' in the human genome according to the evolution-free gospel of ENCODE, *Genome Biology and Evolution* 5: 578–90 (2013), p. 579.

285. Graur, D., Zheng, Y., Price, N., et al., On the immortality of television sets: 'function' in the human genome according to the evolution-free gospel of ENCODE, *Genome Biology and Evolution* 5: 578–90 (2013), p. 587.

286. Top 10 most expensive military planes, *TIME*, <http://content.time.com/time/photogallery/0,29307,1912203_1913325,00.html> (2014).

287. Bhattacharjee, Y., The vigilante. *Science* 343: 1306–9 (2014), p. 1309.

288. Gregory, T. R., *BBC interview with Ewan Birney*, *Genomicron*, <http://www.genomicron.evolverzone.com/2013/04/bbc-interview-with-ewan-birney/> (2013).

289. Mattick, J. S. and Dinger, M. E., The extent of functionality in the human genome. *The HUGO Journal* 7: 1–4 (2013).

290. Mattick, J. S. and Dinger, M. E., The extent of functionality in the human genome. *The HUGO Journal* 7: 1–4 (2013), p. 3.

291. Pheasant, M. and Mattick, J. S., Raising the estimate of functional human sequences. *Genome Research* 17: 1245–53 (2007), p. 1250.

292. Lewis, D., What is our junk DNA for?, *Cosmos Magazine*, <https://cosmosmagazine.com/life-sciences/what-our-junk-dna> (2014).

293. Morris, K. V. and Mattick, J. S., The rise of regulatory RNA. *Nature Reviews Genetics* 15: 423–37 (2014), p. 432.

294. Keller, E. F., From gene action to reactive genomes, *The Journal of Physiology* 592: 2423–9 (2014), p. 2425.

Chapter 7: The Genome in 3D

295. How many chromosomes do people have?, *Genetics Home Reference*, <http://ghr.nlm.nih.gov/handbook/basics/howmanychromosomes> (2014).

296. Yong, E., Getting to Know the Genome, *The Scientist*, <http://www.the-scientist.com/?articles.view/articleNo/32583/title/Getting-to-Know-the-Genome/> (2012).

297. Ayala, F. J., 'Nothing in biology makes sense except in the light of evolution': Theodosius Dobzhansky: 1900–1975. *Journal of Heredity* 68: 3–10 (1977), p. 3.

298. Hall, E., Reading maps of the genes: interpreting the spatiality of genetic knowledge. *Health & Place* 9: 151–61 (2003).

299. Paweletz, N., Walther Flemming: pioneer of mitosis research. *Nature Reviews Molecular and Cellular Biology* 2: 72–5 (2001).

300. Lavelle, C., Pack, unpack, bend, twist, pull, push: the physical side of gene expression. *Current Opinion in Genetics & Development* 25: 74–84 (2014).

301. Cremer, T. and Cremer, M., Chromosome territories. *Cold Spring Harbor Perspectives in Biology* 2: a003889 (2010).

302. Rodriguez, A. and Bjerling, P., The links between chromatin spatial organization and biological function. *Biochemical Society Transactions* 41: 1634–9 (2013).

303. Dahm, R., Friedrich Miescher and the discovery of DNA. *Developmental Biology* 278: 274–88 (2005).

304. Morinière, J., Rousseaux, S., Steuerwald, U., et al. Cooperative binding of two acetylation marks on a histone tail by a single bromodomain. *Nature* 461: 664–8 (2009).

305. Putting the squeeze on sperm DNA: streamlined sperm offer new way to read histone code, *Science Daily*, <http://www.sciencedaily.com/releases/2009/09/090930132652.htm> (2009).

306. Hammoud, S. S., Nix, D. A., Zhang, H., et al., Distinctive chromatin in human sperm packages genes for embryo development. *Nature* 460: 473–8 (2009).

307. Gibcus, J. H. and Dekker, J. The hierarchy of the 3D genome. *Molecular Cell* 49: 773–82 (2013).

308. Bishop, R., Applications of fluorescence *in situ* hybridization (FISH) in detecting genetic aberrations of medical significance. *Bioscience Horizons* 3: 85–95 (2010).

309. Berardi, M. J. and Fantin, V. R., Survival of the fittest: metabolic adaptations in cancer. *Current Opinion in Genetics & Development* 21: 59–66 (2011).

310. Liehr, T., Starke, H., Weise, A., et al., Multicolor FISH probe sets and their applications. *Histology and Histopathology* 19: 229–37 (2004).

311. Klonisch, T., Wark, L., Hombach-Klonisch, S., and Mai, S., Nuclear imaging in three dimensions: a unique tool in cancer research. *Annals of Anatomy = Anatomischer Anzeiger: Official Organ of the Anatomische Gesellschaft* 192: 292–301 (2010).

312. Tchélidzé, P., Chatron-Colliet, A., Thiry, M., et al., Tomography of the cell nucleus using confocal microscopy and medium voltage electron microscopy. *Critical Reviews in Oncology/Hematology* 69: 127–43 (2009).

313. Gorkin, D. U., Leung, D., and Ren, B., The 3D genome in transcriptional regulation and pluripotency. *Cell Stem Cell* 14: 762–75 (2014).

314. Zimmer, M., *Glowing Genes: A Revolution in Biotechnology* (Prometheus Books, 2005).

315. Tsien, R. Y., The green fluorescent protein. *Annual Review of Biochemistry* 67: 509–44 (1998).

316. Tsien, R. Y., Intracellular signal transduction in four dimensions: from molecular design to physiology. *Bowditch Lecture*, C723–C728 (1992).

317. Giepmans, B. N., Adams, S. R., Ellisman, M. H., and Tsien, R. Y., The fluorescent toolbox for assessing protein location and function. *Science* 312: 217–24 (2006).

318. Roger Y. Tsien's speech at the Nobel banquet, *Nobel Media*, <http://www.nobelprize.org/nobel_prizes/chemistry/laureates/2008/tsien-speech_en.html?print=1#.U_0AqMVdXh5> (2008).

319. How bad luck and bad networking cost Douglas Prasher a Nobel Prize, *Discover Magazine*, <http://discovermagazine.com/2011/apr/30-how-bad-luck-networking-cost-prasher-nobel> (2011).

320. Annibale, P. and Gratton, E., Advanced fluorescence microscopy methods for the real-time study of transcription and chromatin dynamics. *Transcription* 5: e28425 (2014).

321. Hou, C. and Corces, V. G., Throwing transcription for a loop: expression of the genome in the 3D nucleus. *Chromosoma* 121: 107–16 (2012).

322. Smallwood, A. and Ren, B., Genome organization and long-range regulation of gene expression by enhancers. *Current Opinion in Cell Biology* 25: 387–94 (2013).

323. Levine, M., Cattoglio, C., and Tjian, R., Looping back to leap forward: transcription enters a new era. *Cell* 157: 13–25 (2014).

324. Lai, F. and Shiekhattar, R., Enhancer RNAs: the new molecules of transcription. *Current Opinion in Genetics & Development* 25: 38–42 (2014).

325. Vance, K. W. and Ponting, C. P., Transcriptional regulatory functions of nuclear long noncoding RNAs. *Trends in Genetics* 30: 348–55 (2014).

326. Mercer, T. R., Edwards, S. L., Clark, M. B., et al., DNase I-hypersensitive exons colocalize with promoters and distal regulatory elements. *Nature Genetics* 45: 852–9 (2013).

327. Heather, A. The genome's 3D structure shapes how genes are expressed, *Garvan Institute*, <http://www.garvan.org.au/news-events/news/the-genome2019s-3d-structure-shapes-how-genes-are-expressed> (2013).

328. Papantonis, A. and Cook, P. R., Transcription factories: genome organization and gene regulation. *Chemical Reviews* 113: 8683–705 (2013).

329. Schoenfelder, S., Sexton, T., Chakalova, L., et al., Preferential associations between co-regulated genes reveal a transcriptional interactome in erythroid cells. *Nature Genetics* 42: 53–61 (2010).

330. Dean, A., In the loop: long range chromatin interactions and gene regulation. *Briefings in Functional Genomics* 10: 3–10 (2011).

331. Strandberg, B., Dickerson, R. E., and Rossmann, M. G., 50 years of protein structure analysis. *Journal of Molecular Biology* 392: 2–32 (2009).

332. Higgs, D. R., Engel, J. D., and Stamatoyannopoulos, G., Thalassaemia. *The Lancet* 379: 373–83 (2012).

333. Ginder, G. D., Epigenetic regulation of fetal globin gene expression in adult erythroid cells. *Translational Research: the journal of laboratory and clinical medicine* (2014).

334. Blanpain, C., Tracing the cellular origin of cancer. *Nature Cell Biology* 15: 126–34 (2013).

335. Nambiar, M., Kari, V., and Raghavan, S. C., Chromosomal translocations in cancer. *Biochimica et Biophysica Acta* 1786: 139–52 (2008).

336. Schwartz, M. and Hakim, O., 3D view of chromosomes, DNA damage, and translocations. *Current Opinion in Genetics & Development* 25: 118–25 (2014).

337. Deng, B., Melnik, S., and Cook, P. R., Transcription factories, chromatin loops, and the dysregulation of gene expression in malignancy. *Seminars in Cancer Biology* 23: 65–71 (2013).

338. Worman, H. J., Ostlund, C., and Wang, Y., Diseases of the nuclear envelope. *Cold Spring Harbor Perspectives in Biology* 2: a000760 (2010).

Chapter 8: The Jumping Genes

339. Keller, E. F., *A Feeling for the Organism: The Life and Work of Barbara McClintock* (W.H. Freeman and Company, 1983), p. 125.

340. Keller, E. F., *A Feeling for the Organism: The Life and Work of Barbara McClintock* (W.H. Freeman and Company, 1983), p. 30.

341. Keller, E. F., *A Feeling for the Organism: The Life and Work of Barbara McClintock* (W.H. Freeman and Company, 1983), p. 55.

342. Zakian, V. A., Telomeres: the beginnings and ends of eukaryotic chromosomes. *Experimental Cell Research* 318: 1456–60 (2012).

343. Jaskelioff, M., Muller, F. L., Paik, J.-H., et al., Telomerase reactivation reverses tissue degeneration in aged telomerase-deficient mice. *Nature* 469: 102–6 (2011).

344. Forêt de Paimpont, *France for Visitors*, <http://france-for-visitors.com/brittany/foret-de-paimpont.html> (2014).

345. Callaway, E., Telomerase reverses ageing process, *Nature News*, <http://www.nature.com/news/2010/101128/full/news.2010.635.html> (2010).

346. Cullen, J. H., *Barbara McClintock* (Chelsea House, 2003), p. 66.

347. Cox, K. H., Bonthuis, P. J., and Rissman, E. F., Mouse model systems to study sex chromosome genes and behavior: relevance to humans. *Frontiers in Neuroendocrinology* 35: 405–19 (2014).

348. Jones, K. T., Meiosis in oocytes: predisposition to aneuploidy and its increased incidence with age. *Human Reproduction Update* 14: 143–58 (2008).

349. Rebollo, R., Romanish, M. T., and Mager, D. L., Transposable elements: an abundant and natural source of regulatory sequences for host genes. *Annual Review of Genetics* 46: 21–42 (2012).

350. Keller, E. F., *A Feeling for the Organism: The Life and Work of Barbara McClintock* (W.H. Freeman and Company, 1983), pp. 175–8.

351. Keller, E. F., *A Feeling for the Organism: The Life and Work of Barbara McClintock* (W.H. Freeman and Company, 1983), p. 142.

352. O'Donnell, K. A. and Burns, K. H., Mobilizing diversity: transposable element insertions in genetic variation and disease. *Mobile DNA* 1: 21 (2010).

353. Ayarpadikannan, S. and Kim, H. S., The impact of transposable elements in genome evolution and genetic instability and their implications in various diseases. *Genomics and Informatics* 12: 98–104 (2014).

354. Weiss, R. A., On the concept and elucidation of endogenous retroviruses. *Philosophical Transactions of the Royal Society of London. Series B, Biological Sciences* 368: 20120494 (2013).

355. Lippincott, S., *David Baltimore: interviewed by Sara Lippincott*, <http://oralhistories.library.caltech.edu/168/1/Baltimore,D._OHO.pdf> (2009).

356. Boeke, J. D., The unusual phylogenetic distribution of retrotransposons: a hypothesis. *Genome Research* 13: 1975–83 (2003).

357. Alzohairy, A. M., Gyulai, G., Jansen, R. K., and Bahieldin, A., Transposable elements domesticated and neofunctionalized by eukaryotic genomes. *Plasmid* 69: 1–15 (2013).

358. Orgel, L. E. and Crick, F. H., Selfish DNA: the ultimate parasite. *Nature* 284: 604–7 (1980), p. 605.

359. Huang, C. R., Burns, K. H., and Boeke, J. D., Active transposition in genomes. *Annual Review of Genetics* 46: 651–75 (2012).

360. Claeys Bouuaert, C., Lipkow, K., Andrews, S. S., et al., The autoregulation of a eukaryotic DNA transposon. *eLife* 2: e00668 (2013).

361. Thorne, E. Why our prehistoric, parasitic 'jumping' genes don't send us into meltdown, *University of Nottingham*, <http://www.nottingham.ac.uk/news/pressreleases/2013/june/why-our-prehistoric,-parasitic-jumping-genes-dont-send-us-into-meltdown.aspx> (2013).

362. Vence, T., 'Sleeping Beauty' named Molecule of the Year, *BioTechniques*, <http://www.biotechniques.com/news/Sleeping-Beauty-named-Molecule-of-theyear/biotechniques-187068.html> (2010).

363. Leslie, M., The immune system's compact genetic counterpart. *Science* 339: 25–7 (2013).

364. Iguchi, Y., Katsuno, M., Ikenaka, K., et al., Amyotrophic lateral sclerosis: an update on recent genetic insights. *Journal of Neurology* 260: 2917–27 (2013).

365. Li, W., Jin, Y., Prazak, L., et al., Transposable elements in TDP-43-mediated neurodegenerative disorders. *PLoS One* 7: e44099 (2012).

366. Storm of 'awakened' transposons may cause brain-cell pathologies in ALS, other illnesses, *Science Daily*, <http://www.sciencedaily.com/releases/2012/09/120906123238.htm> (2012).

367. Dupressoir, A., Lavialle, C., and Heidmann, T., From ancestral infectious retroviruses to bona fide cellular genes: role of the captured syncytins in placentation. *Placenta* 33: 663–71 (2012), p. 663.

368. Halaby, D. M., Poupon, A., and Mornon, J. P., The immunoglobulin fold family: sequence analysis and 3D structure comparisons. *Protein Engineering* 12: 563–71 (1999).

369. Fugmann, S. D., The origins of the Rag genes—from transposition to V(D)J recombination. *Seminars in Immunology* 22: 10–16 (2010).

370. Reilly, M. T., Faulkner, G. J., Dubnau, J., et al., The role of transposable elements in health and diseases of the central nervous system. *The Journal of Neuroscience* 33: 17577–86 (2013).

371. Lupski, J. R., Genetics: genome mosaicism—one human, multiple genomes. *Science* 341: 358–9 (2013).

372. Platt, R. N. 2nd, Vandewege, M. W., Kern, C., et al. Large numbers of novel miRNAs originate from DNA transposons and are coincident with a large species radiation in bats. *Molecular Biology and Evolution* 31: 1536–45 (2014).

373. *Science Daily*, <http://www.sciencedaily.com/releases/2014/04/140401173134.htm> (2014).

374. Stindl, R., The telomeric sync model of speciation: species-wide telomere erosion triggers cycles of transposon-mediated genomic rearrangements, which underlie the saltatory appearance of nonadaptive characters. *Die Naturwissenschaften* 101: 163–86 (2014).

375. Britten, R. J., Transposable element insertions have strongly affected human evolution. *Proceedings of the National Academy of Sciences of the United States of America* 107: 19945–8 (2010).

376. Jorgensen, R. A., Restructuring the genome in response to adaptive challenge: McClintock's bold conjecture revisited. *Cold Spring Harbor Symposia on Quantitative Biology* 69: 349–54 (2004), p. 349.

Chapter 9: The Marks of Lamarck

377. Bard, J. B., The next evolutionary synthesis: from Lamarck and Darwin to genomic variation and systems biology. *Cell Communication and Signaling: CCS* 9: 30 (2011).

378. Honeywill, R., *Lamarck's Evolution: Two Centuries of Genius and Jealousy* (Murdoch Books, 2008), p. 6.

379. Darwin, C., *Origin of Species* (John Murray, 1869). p. xv.

380. Cuvier, G., Elegy of Lamarck, *The Victorian Web*, <http://www.victorianweb.org/science/science_texts/cuvier/cuvier_on_lamarck.htm> (2014).

381. Handel, A. E. and Ramagopalan, S. V., Is Lamarckian evolution relevant to medicine? *BMC Medical Genetics* 11: 73 (2010).

382. Bowler, P. J., *Evolution: The History of an Idea* (University of California Press, 2003).

383. Cavalier-Smith, T., Cell evolution and Earth history: stasis and revolution. *Philosophical Transactions of the Royal Society of London. Series B, Biological Sciences* 361: 969–1006 (2006).

384. Desmond, A. and Moore, J. R., *Darwin* (Michael Joseph, 1991), p. 286.

385. Shermer, M., *In Darwin's Shadow* (Oxford University Press, 2002), p. 45.

386. Koonin, E. V., Calorie restriction a Lamarck. *Cell* 158: 237–8 (2014), p. 238.

387. Slack, J. M., Conrad Hal Waddington: the last Renaissance biologist? *Nature Reviews Genetics* 3: 889–95 (2002).

388. Badano, J. L. and Katsanis, N., Beyond Mendel: an evolving view of human genetic disease transmission. *Nature Reviews Genetics* 3: 779–89 (2002).

389. Berridge, M. J., Bootman, M. D., and Roderick, H. L., Calcium signalling: dynamics, homeostasis and remodelling. *Nature Reviews Molecular Cell Biology* 4: 517–29 (2003).

390. Roseboom, T. J., van der Meulen, J. H., Ravelli, A. C., et al., Effects of prenatal exposure to the Dutch famine on adult disease in later life: an overview. *Molecular and Cellular Endocrinology* 185: 93–8 (2001).

391. Carey, N., *Beyond DNA: epigenetics*, http://www.naturalhistorymag.com/features/142195/beyond-dna-epigenetics (2012).

392. Pembrey, M. E., Male-line transgenerational responses in humans. *Human Fertility* 13: 268–71 (2010).

393. Choudhuri, S., From Waddington's epigenetic landscape to small noncoding RNA: some important milestones in the history of epigenetics research. *Toxicology Mechanisms and Methods* 21: 252–74 (2011).

394. Rothbart, S. B. and Strahl, B. D., Interpreting the language of histone and DNA modifications. *Biochimica et Biophysica Acta* 1839: 627–43 (2014).

395. Nakagawa, S. and Kageyama, Y., Nuclear lncRNAs as epigenetic regulators—beyond skepticism. *Biochimica et Biophysica Acta* 1839: 215–22 (2014).

396. Boland, M. J., Nazor, K. L., and Loring, J. F., Epigenetic regulation of pluripotency and differentiation. *Circulation Research* 115: 311–24 (2014).

397. Rothbart, S. B. and Strahl, B. D., Interpreting the language of histone and DNA modifications. *Biochimica et Biophysica Acta* 1839: 627–43 (2014).

398. Peters, J., The role of genomic imprinting in biology and disease: an expanding view. *Nature Reviews Genetics* 15: 517–30 (2014).

399. Zucchi, F. C., Yao, Y., and Metz, G. A., The secret language of destiny: stress imprinting and transgenerational origins of disease. *Frontiers in Genetics* 3: 96 (2012).

400. Narbonne, P., Miyamoto, K., and Gurdon, J. B., Reprogramming and development in nuclear transfer embryos and in interspecific systems. *Current Opinion in Genetics & Development* 22: 450–8 (2012).

401. Tammen, S. A., Friso, S., and Choi, S. W., Epigenetics: the link between nature and nurture. *Molecular Aspects of Medicine* 34: 753–64 (2013).

402. Griffiths, B. B. and Hunter, R. G., Neuroepigenetics of stress. *Neuroscience* 275C: 420–35 (2014).

403. Mitchell, C., Hobcraft, J., McLanahan, S. S., et al., Social disadvantage, genetic sensitivity, and children's telomere length. *Proceedings of the National Academy of Sciences of the United States of America* 111: 5944–9 (2014).

404. Campos, E. I., Stafford, J. M., and Reinberg, D., Epigenetic inheritance: histone bookmarks across generations. *Trends in Cell Biology* 24: 664–74 (2014).

405. Hughes, V., The sins of the father. *Nature* 507: 22–4 (2014).

406. Gapp, K., Jawaid, A., Sarkies, P., et al., Implication of sperm RNAs in transgenerational inheritance of the effects of early trauma in mice. *Nature Neuroscience* 17: 667–9 (2014).

407. Smythies, J., Edelstein, L., and Ramachandran, V., Molecular mechanisms for the inheritance of acquired characteristics-exosomes, microRNA shuttling, fear and stress: Lamarck resurrected? *Frontiers in Genetics* 5: 133 (2014).

408. Koonin, E. V., Calorie restriction a Lamarck. *Cell* 158: 237–8 (2014), p. 238.

409. Koonin, E. V., Calorie restriction a Lamarck. *Cell* 158: 237–8 (2014), p. 237.

410. Stindl, R., The telomeric sync model of speciation: species-wide telomere erosion triggers cycles of transposon-mediated genomic rearrangements, which underlie the saltatory appearance of nonadaptive characters. *Die Naturwissenschaften* 101: 163–86 (2014).

411. Stindl, R., The telomeric sync model of speciation: species-wide telomere erosion triggers cycles of transposon-mediated genomic rearrangements, which underlie the saltatory appearance of nonadaptive characters. *Die Naturwissenschaften* 101: 163–86 (2014), p. 173.

412. Stindl, R., The telomeric sync model of speciation: species-wide telomere erosion triggers cycles of transposon-mediated genomic rearrangements, which underlie the saltatory appearance of nonadaptive characters. *Die Naturwissenschaften* 101: 163–86 (2014), p. 176.

Chapter 10: Code, Non-Code, Garbage, and Junk

413. Jayaraman, R., Jacques Monod and the advent of the age of operons. *Resonance* 15: 1084–96 (2010), p. 1084.

414. Kellis, M., Wold, B., Snyder, M. P., et al., Defining functional DNA elements in the human genome. *Proceedings of the National Academy of Sciences of the United States of America* 111: 6131–8 (2014).

415. Haerty, W. and Ponting, C. P., No gene in the genome makes sense except in the light of evolution. *Annual Review of Genomics and Human Genetics* 15: 71–92 (2014), p. 72.

416. Haerty, W. and Ponting, C. P., No gene in the genome makes sense except in the light of evolution. *Annual Review of Genomics and Human Genetics* 15: 71–92 (2014).

417. Haerty, W. and Ponting, C. P., No gene in the genome makes sense except in the light of evolution. *Annual Review of Genomics and Human Genetics* 15: 71–92 (2014), p. 73.

418. Rands, C. M., Meader, S., Ponting, C. P., and Lunter, G., 8.2% of the human genome is constrained: variation in rates of turnover across functional element classes in the human lineage. *PLoS Genetics* 10: e1004525 (2014).

419. Pheasant, M. and Mattick, J. S., Raising the estimate of functional human sequences. *Genome Research* 17: 1245–53 (2007).

420. Mattick, J. S. and Dinger, M. E., The extent of functionality in the human genome. *The HUGO Journal* 7: 2 (2013).

421. Roberts, J. T., Cardin, S. E., and Borchert, G. M., Burgeoning evidence indicates that microRNAs were initially formed from transposable element sequences. *Mobile Genetic Elements* 4: e29255 (2014).

422. Stamatoyannopoulos, J. A., What does our genome encode? *Genome Research* 22: 1602–11 (2012).

423. Palazzo, A., Junk DNA—origin of the term, *Science Blogs*, <http://scienceblogs.com/transcript/2007/02/12/junk-dna-origin-of-the-term-1/> (2007).

424. Ohno, S., So much 'junk' DNA in our genome. *Brookhaven Symposium on Biology* 23: 366–70 (1972).

425. Leslie, M., 'Dead' enzymes show signs of life. *Science* 340: 25–7 (2013).

426. Leslie, M., 'Dead' enzymes show signs of life. *Science* 340: 25–7 (2013), p. 25.

427. Thomas, C. A., The genetic organization of chromosomes. *Annual Review of Genetics* 5: 237–56 (1971).

428. Doolittle, W. F., Is junk DNA bunk? A critique of ENCODE. *Proceedings of the National Academy of Sciences of the United States of America* 110: 5294–300 (2013), p. 5295.

429. Buerk, R., Fugu: the fish more poisonous than cyanide, *BBC*, <http://www.bbc.co.uk/news/magazine-18065372> (2012).

430. Mattick, J. S. and Dinger, M. E., The extent of functionality in the human genome. *The HUGO Journal* 7: 2 (2013).

431. Lewis, D., What is our junk DNA for?, *Cosmos Magazine*, <https://cosmosmagazine.com/life-sciences/what-our-junk-dna> (2014).

432. Skloot, R., *The Immortal Life of Henrietta Lacks* (Macmillan, 2010).

433. *Henrietta Lacks: family win recognition for immortal cells*, *BBC*, <http://www.bbc.co.uk/news/world-us-canada-23611189> (2013).

434. Stamatoyannopoulos, J. A., What does our genome encode? *Genome Research* 22: 1602–11 (2012), p. 1609.

435. Mattick, J. S. and Dinger, M. E., The extent of functionality in the human genome. *The HUGO Journal* 7: 2 (2013).

436. Badano, J. L. and Katsanis, N., Beyond Mendel: an evolving view of human genetic disease transmission. *Nature Reviews Genetics* 3: 779–89 (2002).

437. Gupta, R. M. and Musunuru, K., Expanding the genetic editing tool kit: ZFNs, TALENs, and CRISPR-Cas9. *Journal of Clinical Investigation* 124: 4154–61 (2014).

438. Bassett, A. R., Azzam, G., Wheatley, L., et al., Understanding functional miRNA-target interactions in vivo by site-specific genome engineering. *Nature Communications* 5: 4640 (2014).

439. Bassett, A. R., Akhtar, A., Barlow, D. P., et al., Considerations when investigating lncRNA function in vivo. *eLife* 3: e03058 (2014).

440. Meier, I. D., Bernreuther, C., Tilling, T., et al., Short DNA sequences inserted for gene targeting can accidentally interfere with off-target gene expression. *FASEB Journal: Official Publication of the Federation of American Societies for Experimental Biology* 24: 1714–24 (2010).

441. Cathomen, T. and Ehl, S., Translating the genomic revolution—targeted genome editing in primates. *New England Journal of Medicine* 370: 2342–5 (2014).

442. Niu, J., Zhang, B., and Chen, H., Applications of TALENs and CRISPR/Cas9 in human cells and their potentials for gene therapy. *Molecular Biotechnology* 56: 681–8 (2014).

Chapter 11: Genes and Disease

443. Scriver, C. R., Garrod's foresight; our hindsight. *Journal of Inherited Metabolic Disease* 24: 93–116 (2001).

444. Cox, K. H., Bonthuis, P. J., and Rissman, E. F., Mouse model systems to study sex chromosome genes and behavior: relevance to humans. *Frontiers in Neuroendocrinology* 35: 405–19 (2014).

445. Deng, X., Berletch, J. B., Nguyen, D. K., and Disteche, C. M., X chromosome regulation: diverse patterns in development, tissues and disease. *Nature Reviews Genetics* 15: 367–78 (2014).

446. Schramm, W., The history of haemophilia—a short review. *Thrombosis Research* 134S1: S4-S9 (2014).

447. Lannoy, N. and Hermans, C., The 'royal disease'—haemophilia A or B? A haematological mystery is finally solved. *Haemophilia: The Official Journal of the World Federation of Hemophilia* 16: 843–7 (2010).

448. Reed, W. and Vichinsky, E. P., New considerations in the treatment of sickle cell disease. *Annual Review of Medicine* 49: 461–74 (1998).

449. Higgs, D. R., Engel, J. D., and Stamatoyannopoulos, G., Thalassaemia. *The Lancet* 379: 373–83 (2012).

450. It's in the Blood!, *Oregon State University*, <http://scarc.library.oregonstate.edu/coll/pauling/blood/quotes/linus_pauling.html> (2014).

451. Kim, S. D. and Fung, V. S., An update on Huntington's disease: from the gene to the clinic. *Current Opinion in Neurology* 27: 477–83 (2014).

452. Wild, E., Interview: Alice and Nancy Wexler, *HDBuzz*, <http://en.hdbuzz.net/101> (2012).

453. Charlotte Raven: should I take my own life?, *The Guardian*, <http://www.theguardian.com/society/2010/jan/16/charlotte-raven-should-i-take-my-own-life> (2010).

454. Quinton, P. M., Cystic fibrosis: lessons from the sweat gland. *Physiology* 22: 212–25 (2007), p. 212.

455. Elmer-Dewitt, P., The Genetic Revolution, *TIME*: 46–50 (1994), p. 48.

456. Rahimov, F. and Kunkel, L. M., The cell biology of disease: cellular and molecular mechanisms underlying muscular dystrophy. *The Journal of Cell Biology* 201: 499–510 (2013).

457. Taussig, N., Our beautiful sons could die before us, *The Guardian*, <http://www.theguardian.com/lifeandstyle/2014/aug/16/our-beautiful-sons-could-die-before-us> (2014).

458. Pearson, H., Human Genetics: One Gene, Twenty Years, *Nature News*, <http://www.nature.com/news/2009/090708/full/460164a.html> (2009), p.165.

459. Scriver, C. R., Garrod's foresight; our hindsight. *Journal of Inherited Metabolic Disease* 24: 93–116 (2001).

460. Phenylketonuria, *NHS*, <http://www.nhs.uk/conditions/phenylketonuria/Pages/Introduction.aspx> (2014).

461. Touzot, F., Hacein-Bey-Abina, S., Fischer, A., and Cavazzana, M., Gene therapy for inherited immunodeficiency. *Expert Opinion on Biological Therapy* 14: 789–98 (2014).

462. Chen, H., Ruan, Y. C., Xu, W. M., et al., Regulation of male fertility by CFTR and implications in male infertility. *Human Reproduction Update* 18: 703–13 (2012).

463. Gallati, S., Disease-modifying genes and monogenic disorders: experience in cystic fibrosis. *The Application of Clinical Genetics* 7: 133–46 (2014).

464. Waalen, J. and Beutler, E., Genetic screening for low-penetrance variants in protein-coding genes. *Annual Review of Genomics and Human Genetics* 10: 431–50 (2009).

465. What they said: genome in quotes, *BBC*, <http://news.bbc.co.uk/1/hi/sci/tech/807126.stm> (2000).

466. Rose, H. and Rose, S., *Genes, Cells and Brains* (Verso, 2012), p. 25.

467. Genome announcement a milestone, but only a beginning, *CNN*, <http://edition.cnn.com/2000/HEALTH/06/26/human.genome.05/> (2000).

468. Kang, B., Park, J., Cho, S., et al., Current status, challenges, policies, and bioethics of biobanks. *Genomics & Informatics* 11: 211–17 (2013).

469. Nakamura, Y., DNA variations in human and medical genetics: 25 years of my experience. *Journal of Human Genetics* 54: 1–8 (2009).

470. Visscher, P. M., Brown, M. A., McCarthy, M. I., and Yang, J., Five years of GWAS discovery. *American Journal of Human Genetics* 90: 7–24 (2012).

471. Grarup, N., Sandholt, C. H., Hansen, T., and Pedersen, O., Genetic susceptibility to type 2 diabetes and obesity: from genome-wide association studies to rare variants and beyond. *Diabetologia* 57: 1528–41 (2014).

472. Wallberg, M. and Cooke, A., Immune mechanisms in type 1 diabetes. *Trends in Immunology* 34: 583–91 (2013).

473. Visscher, P. M., Brown, M. A., McCarthy, M. I., and Yang, J., Five years of GWAS discovery. *American Journal of Human Genetics* 90: 7–24 (2012), p. 7.

474. McClellan, J. and King, M. C., Genetic heterogeneity in human disease. *Cell* 141: 210–17 (2010), p. 213.

475. McClellan, J. and King, M. C., Genetic heterogeneity in human disease. *Cell* 141: 210–17 (2010), p. 216.

476. Mundasad, S., 'Eighty new genes linked to schizophrenia', *BBC*, <http://www.bbc.co.uk/news/health-28401693> (2014).

477. Stamatoyannopoulos, J. A., What does our genome encode? *Genome Research* 22: 1602–11 (2012).

478. Schizophrenia Working Group of the Psychiatric Genomics Consortium, Biological insights from 108 schizophrenia-associated genetic loci. *Nature* 511: 421–7 (2014).

479. Clarke, L., Ten years on: the abiding presence of R. D. Laing. *Journal of Psychiatric and Mental Health Nursing* 6: 313–20 (1999).

480. Laing, R. D., *The Divided Self: An Existential Study in Sanity and Madness* (Penguin, 1969), p. 12.

481. Sherwell, P., DNA father James Watson's 'holy grail' request, *The Telegraph*, <http://www.telegraph.co.uk/news/worldnews/northamerica/usa/5300883/DNA-father-James-Watsons-holy-grail-request.html> (2009).

482. Pinto, R., Ashworth, M., and Jones, R., Schizophrenia in black Caribbeans living in the UK: an exploration of underlying causes of the high incidence rate. *The British Journal of General Practice: The Journal of the Royal College of General Practitioners* 58: 429–34 (2008).

483. Pinto, R., Ashworth, M., and Jones, R., Schizophrenia in black Caribbeans living in the UK: an exploration of underlying causes of the high incidence rate. *The British Journal of General Practice: The Journal of the Royal College of General Practitioners* 58: 429–34 (2008), p. 433.

484. Singh, S., Kumar, A., Agarwal, S., et al., Genetic insight of schizophrenia: past and future perspectives. *Gene* 535: 97–100 (2014), p. 97.

485. Uher, R., Gene-environment interactions in severe mental illness. *Frontiers in Psychiatry* 5: 48 (2014).

486. SPIEGEL interview with Craig Venter: 'We Have Learned Nothing from the Genome', *Der Spiegel*, <http://www.spiegel.de/international/world/spiegel-interview-with-craig-venter-we-have-learned-nothing-from-the-genome-a-709174-2.html> (2010).

487. Saint Pierre, A. and Genine, E., How important are rare variants in common disease? *Briefings in Functional Genomics* 13: 353-61 (2014).

488. Panoutsopoulou, K., Tachmazidou, I., and Zeggini, E., In search of low-frequency and rare variants affecting complex traits. *Human Molecular Genetics* 22: R16–21 (2013).

489. Schreiber, M., Dorschner, M., and Tsuang, D., Next-generation sequencing in schizophrenia and other neuropsychiatric disorders. *American Journal of Medical Genetics. Part B, Neuropsychiatric Genetics: The Official Publication of the International Society of Psychiatric Genetics* 162B: 671–8 (2013).

490. Purcell, S. M., Moran, J. L., Fromer, M., et al., A polygenic burden of rare disruptive mutations in schizophrenia. *Nature* 506: 185–90 (2014).

491. Schreiber, M., Dorschner, M., and Tsuang, D., Next-generation sequencing in schizophrenia and other neuropsychiatric disorders. *American Journal of Medical Genetics. Part B, Neuropsychiatric Genetics: The Official Publication of the International Society of Psychiatric Genetics* 162B: 671–8 (2013), p. 672.

492. Li, C., Personalized medicine—the promised land: are we there yet? *Clinical Genetics* 79: 403–12 (2011).

493. St Laurent, G., Vyatkin, Y., and Kapranov, P., Dark matter RNA illuminates the puzzle of genome-wide association studies. *BMC Medicine* 12: 97 (2014).

494. Sadee, W., Hartmann, K., Seweryn, M., Pietrzak, M., Handelman, S. K., and Rempala, G. A. Missing heritability of common diseases and treatments outside the protein-coding exome. *Human Genetics*. 133: 1199–215 (2014).

Chapter 12: What Makes Us Human?

495. Darwin, C., *Origin of Species* (John Murray, 1859), p. 488.

496. Shermer, M., *In Darwin's Shadow* (Oxford University Press, 2002), p. 161.

497. Suddendorf, T., *The Gap* (Basic Books, 2013), p. 8.

498. Spencer, G., New Genome Comparison Finds Chimps, Humans Very Similar at the DNA Level, *National Human Genome Research Institute*, <http://www.genome.gov/15515096> (2005).

499. Why mouse matters, *National Human Genome Research Institute*, <http://www.genome.gov/ 10001345> (2010).

500. Swanson, K. S., Mazur, M. J., Vashisht, K., et al., Genomics and clinical medicine: rationale for creating and effectively evaluating animal models. *Experimental Biology and Medicine* 229: 866–75 (2004).

501. Debiec, J., Peptides of love and fear: vasopressin and oxytocin modulate the integration of information in the amygdala. *BioEssays: News and Reviews in Molecular, Cellular and Developmental Biology* 27: 869–73 (2005).

502. McGrew, W. C., Is primate tool use special? Chimpanzee and New Caledonian crow compared. *Philosophical Transactions of the Royal Society of London. Series B, Biological Sciences* 368: 20120422 (2013).

503. Suddendorf, T., *The Gap* (Basic Books, 2013), p. 65.

504. Suddendorf, T., *The Gap* (Basic Books, 2013), p. 74.

505. Kuhl, P. K., Brain mechanisms in early language acquisition. *Neuron* 67: 713–27 (2010), p. 713.

506. Trigger, B., Comment on Tobias, Piltdown, the case against Keith. *Current Anthropology* 33: 275 (1992).

507. Engels, F., The part played by labour in the transition from ape to man, *Marxists.org*, <http://www.marxists.org/archive/marx/works/1876/part-played-labour/> (1876).

508. Sheehan, H., *Marxism and the Philosophy of Science: A Critical History* (Humanities Press International, 1993), p. 24.

509. Stringer, C., *The Origin of our Species* (Allen Lane, 2011), p. 213.

510. Corballis, M., Not the last word, *American Scientist*, <http://www.americanscientist.org/ bookshelf/pub/not-the-last-word> (2007).

511. Stringer, C., *The Origin of our Species* (Allen Lane, 2011), p. 158.

512. Suddendorf, T., *The Gap* (Basic Books, 2013), p. 76.

513. Dunbar, R., *Grooming, Gossip and the Evolution of Language* (Faber and Faber, 2004).

514. Deacon, T. W., *The Symbolic Species: The Co-evolution of Language and the Brain* (W. W. Norton & Company, 1998).

515. Van der Veer, R. and Valsiner, J., *Understanding Vygotsky: A Quest for Synthesis* (Blackwell, 1991).

516. Vygotsky, L. S., *Mind in Society* (Harvard University Press, 1978), p. 55.

517. Kozulin, A., *Vygotsky's Psychology* (Harvester Wheatsheaf, 1990), p. 225.

518. Callaway, E., 'Smart genes' prove elusive, *Nature News*, <http://www.nature.com/news/ smart-genes-prove-elusive-1.15858> (2014).

519. Darwin's Tree of Life, *Natural History Museum*, <http://www.nhm.ac.uk/nature-online/ evolution/tree-of-life/darwin-tree/> (2014).

520. Stringer, C., *The Origin of our Species* (Allen Lane, 2011), p. 266.

521. Stringer, C., *The Origin of our Species* (Allen Lane, 2011), p. 124.

522. Pääbo, S., *Neanderthal Man* (Basic Books, 2014).

523. Russell, O. and Turnbull, D., Mitochondrial DNA disease-molecular insights and potential routes to a cure. *Experimental Cell Research* 325: 38–43 (2014).

524. Rose, S., Lynn Margulis obituary, *The Guardian*, <http://www.theguardian.com/science/ 2011/dec/11/lynn-margulis-obtiuary> (2011).

525. Wong, L. J., Pathogenic mitochondrial DNA mutations in protein-coding genes. *Muscle & Nerve* 36: 279–93 (2007).

526. Sato, M. and Sato, K., Maternal inheritance of mitochondrial DNA by diverse mechanisms to eliminate paternal mitochondrial DNA. *Biochimica et Biophysica Acta* 1833: 1979–84 (2013).

527. Stringer, C., *The Origin of our Species* (Allen Lane, 2011), p. 23.

528. Stringer, C., *The Origin of our Species* (Allen Lane, 2011), p. 171.

529. Golding, W., *The Inheritors* (Faber and Faber, 1955).

530. Kolbert, E., *The Sixth Extinction* (Henry Holt & Company, 2014).

531. Stringer, C., *The Origin of our Species* (Allen Lane, 2011), p. 52.

532. Higham, T., Douka, K., Wood, R., et al., The timing and spatiotemporal patterning of Neanderthal disappearance. *Nature* 512: 306–9 (2014).

533. Pääbo, S., *Neanderthal Man* (Basic Books, 2014), p. 200.

534. Pääbo, S., *Neanderthal Man* (Basic Books, 2014), p. 243.

535. Pääbo, S., *Neanderthal Man* (Basic Books, 2014), p. 176.

536. Vernot, B. and Akey, J. M., Resurrecting surviving Neandertal lineages from modern human genomes. *Science* 343: 1017–21 (2014).

537. Sankararaman, S., Mallick, S., Dannemann, M., et al., The genomic landscape of Neanderthal ancestry in present-day humans. *Nature* 507: 354–7 (2014).

538. Pääbo, S., *Neanderthal Man* (Basic Books, 2014), p. 188.

539. Huerta-Sánchez, E., Jin, X., Asan, et al., Altitude adaptation in Tibetans caused by introgression of Denisovan-like DNA. *Nature* 512: 194–7 (2014).

540. Callaway, E., Modern human genomes reveal our inner Neanderthal, *Nature News*, <http://www.nature.com/news/modern-human-genomes-reveal-our-inner-neanderthal-1.14615> (2014).

541. Bowden, R. MacFie, T. S., Myers, S., et al. Genomic tools for evolution and conservation in the chimpanzee: Pan troglodytes ellioti is a genetically distinct population. *PLoS Genetics* 8: e1002504 (2012).

542. Stringer, C., *The Origin of our Species* (Allen Lane, 2011), p. 126.

543. Kaplan, M., Neanderthals ate their greens, *Nature News*, <http://www.nature.com/news/neanderthals-ate-their-greens-1.11030> (2012).

544. Appenzeller, T., Neanderthal culture: old masters. *Nature* 497: 302–4 (2013), p. 303.

545. Oksenberg, N., Stevison, L., Wall, J. D., and Ahituv, N., Function and regulation of AUTS2, a gene implicated in autism and human evolution. *PLoS Genetics* 9: e1003221 (2013).

546. Ahmed, M. and Liang, P., Study of modern human evolution via comparative analysis with the Neanderthal genome. *Genomics & Informatics* 11: 230–8 (2013).

547. Gokhman, D., Lavi, E., Prüfer, K., et al., Reconstructing the DNA methylation maps of the Neandertal and the Denisovan. *Science* 344: 523–7 (2014).

548. Pennisi, E., Ancient DNA holds clues to gene activity in extinct humans. *Science* 344: 245–6 (2014), p. 246.

549. Barras, C., Why we get autism but our Neanderthal cousins didn't, *New Scientist*, <http://www.newscientist.com/article/dn25443-why-we-get-autism-but-our-neanderthal-cousins-didnt.html#.U_X2zI10wuQ> (2014).

550. Seldon, H. L., Extended neocortical maturation time encompasses speciation, fatty acid and lateralization theories of the evolution of schizophrenia and creativity. *Medical Hypotheses* 69: 1085–9 (2007).

551. De Manzano, O., Cervenka, S., Karabanov, A., et al., Thinking outside a less intact box: thalamic dopamine D2 receptor densities are negatively related to psychometric creativity in healthy individuals. *PLoS One* 5: e10670 (2010).

552. Roberts, M., Creative minds 'mimic schizophrenia', *BBC*, <http://www.bbc.co.uk/news/10154775> (2010).

553. Kavanagh, D. H., Dwyer, S., O'Donovan, M. C., and Owen, M. J., The ENCODE project: implications for psychiatric genetics. *Molecular Psychiatry* 18: 540–2 (2013).

Chapter 13: The Genome That Became Conscious

554. Somel, M., Liu, X., and Khaitovich, P., Human brain evolution: transcripts, metabolites and their regulators. *Nature Reviews Neuroscience* 14: 112–27 (2013).

555. Ahmed, M. and Liang, P., Study of modern human evolution via comparative analysis with the Neanderthal genome. *Genomics & Informatics* 11: 230–8 (2013).

556. Konopka, G., Friedrich, T., Davis-Turak, J., et al., Human-specific transcriptional networks in the brain. *Neuron* 75: 601–17 (2012).

557. Pinker, S., *How the Mind Works* (Allen Lane, 1997), p. 27.

558. Antzoulatos, E. G. and Miller, E. K., Increases in functional connectivity between prefrontal cortex and striatum during category learning. *Neuron* 83: 216–25 (2014).

559. Synchronized brain waves enable rapid learning, *Science Daily*, <http://www.sciencedaily.com/releases/2014/06/140612121354.htm> (2012).

560. Tsien, R. Y., Intracellular signal transduction in four dimensions: from molecular design to physiology. *Bowditch Lecture*, C723–C728 (1992).

561. Leybaert, L. and Sanderson, M. J., Intercellular Ca(2+) waves: mechanisms and function. *Physiological Reviews* 92: 1359–92 (2012).

562. Robison, A. J., Emerging role of CaMKII in neuropsychiatric disease. *Trends in Neurosciences* 37: 653–62 (2014).

563. Schmidt, E., *More sophisticated wiring, not just a bigger brain, helped humans evolve beyond chimps*, <http://newsroom.ucla.edu/releases/more-sophisticated-wiring-not-237689> (2012).

564. Toga, A. W., Thompson, P. M., and Sowell, E. R., Mapping brain maturation. *Trends in Neurosciences* 29: 148–59 (2006).

565. Pievani, T., Many ways of being human, the Stephen J. Gould's legacy to palaeoanthropology (2002–2012). *Journal of Anthropological Sciences = Rivista di antropologia: JASS/Istituto italiano di antropologia* 90: 133–49 (2012).

566. Kelley, J., Neanderthal teeth lined up. *Nature* 428: 904–5 (2004).

567. Orban, G. A. and Caruana, F., The neural basis of human tool use. *Frontiers in Psychology* 5: 310 (2014).

568. Liu, X., Somel, M., Tang, L., et al., Extension of cortical synaptic development distinguishes humans from chimpanzees and macaques. *Genome Research* 22: 611–22 (2012).

569. Menet, J. S. and Rosbash, M., When brain clocks lose track of time: cause or consequence of neuropsychiatric disorders. *Current Opinion in Neurobiology* 21: 849–57 (2011).

570. Liu, C., Teng, Z.-Q., Santistevan, N. J., et al., Epigenetic regulation of miR-184 by MBD1 governs neural stem cell proliferation and differentiation. *Cell Stem Cell* 6: 433–44 (2010).

571. Braun, S. M. and Jessberger, S., Adult neurogenesis: mechanisms and functional significance. *Development* 141: 1983–6 (2014).

572. Barry, G. and Mattick, J. S., The role of regulatory RNA in cognitive evolution. *Trends in Cognitive Sciences* 16: 497–503 (2012), p. 499.

573. Barry, G. and Mattick, J. S., The role of regulatory RNA in cognitive evolution. *Trends in Cognitive Sciences* 16: 497–503 (2012), p. 500.

574. Smalheiser, N. R., The RNA-centred view of the synapse: non-coding RNAs and synaptic plasticity. *Philosophical Transactions of the Royal Society of London. Series B, Biological Sciences* 369: pii: 20130504 (2014).

575. Iyengar, B. R., Choudhary, A., Sarangdhar, M. A., et al., Non-coding RNA interact to regulate neuronal development and function. *Frontiers in Cellular Neuroscience* 8: 47 (2014).

576. Boland, M. J., Nazor, K. L., and Loring, J. F., Epigenetic regulation of pluripotency and differentiation. *Circulation Research* 115: 311–24 (2014).

577. Perrimon, N., Pitsouli, C., and Shilo, B. Z., Signaling mechanisms controlling cell fate and embryonic patterning. *Cold Spring Harbor Perspectives in Biology* 4: a005975 (2012).

578. Rothbart, S. B. and Strahl, B. D., Interpreting the language of histone and DNA modifications. *Biochimica et Biophysica Acta* 1839: 627–43 (2014).

579. Rudenko, A. and Tsai, L. H., Epigenetic regulation in memory and cognitive disorders. *Neuroscience* 264: 51–63 (2014), p. 51.

580. Rudenko, A. and Tsai, L. H., Epigenetic regulation in memory and cognitive disorders. *Neuroscience* 264: 51–63 (2014).

581. Guan, J. S., Xie, H., and Ding, X., The role of epigenetic regulation in learning and memory. *Experimental Neurology* pii: S0014-4886(14)00147-2 (2014).

582. Reilly, M. T., Faulkner, G. J., Dubnau, J., et al., The role of transposable elements in health and diseases of the central nervous system. *The Journal of Neuroscience* 33: 17577–86 (2013).

583. Stringer, C., *The Origin of our Species* (Allen Lane, 2011), p. 116.

584. Barry, G. and Mattick, J. S., The role of regulatory RNA in cognitive evolution. *Trends in Cognitive Sciences* 16: 497–503 (2012), p. 501.

585. Stanford, P. K., August Weismann's theory of the germ-plasm and the problem of unconceived alternatives. *History and Philosophy of the Life Sciences* 27: 163 (2005).

586. Gapp, K., Jawaid, A., Sarkies, P., et al., Implication of sperm RNAs in transgenerational inheritance of the effects of early trauma in mice. *Nature Neuroscience* 17: 667–9 (2014).

587. Hurley, D., Grandma's experiences leave a mark on your genes, *Discover Magazine*, <http://discovermagazine.com/2013/may/13-grandmas-experiences-leave-epigenetic-mark-on-your-genes> (2013).

588. Heard, E. and Martienssen, R. A., Transgenerational epigenetic inheritance: myths and mechanisms. *Cell* 157: 95–109 (2014), p. 95.

589. Heard, E. and Martienssen, R. A., Transgenerational epigenetic inheritance: myths and mechanisms. *Cell* 157: 95–109 (2014), p. 106.

590. Hughes, V., The sins of the father. *Nature* 507: 22–4 (2014), p. 24.

Conclusion: The Case for Complexity

591. Robertson, J. M., *The Philosophical Works of Francis Bacon* (Routledge, 2013), p. 429.

592. Van Regenmortel, M. H. V., Reductionism and complexity in molecular biology. *EMBO Reports* 5: 1016–20 (2004), p. 1016.

593. Boss, M. and Poggi, S., *Romanticism in Science: Science in Europe, 1790–1840* (Springer, 1993), p. 63.

594. Shanahan, T., *The Evolution of Darwinism: Selection, Adaptation and Progress in Evolutionary Biology* (Cambridge University Press, 2004), p. 2.

595. Anderson, A., A Time for Adaptation, *Huffington Post*, <http://www.huffingtonpost.com/anthony-anderson/a-time-for-adaptation_b_170948.html> (2011).

596. Herbert Spencer, *American Experience*, <http://www.pbs.org/wgbh/amex/carnegie/peopleevents/pande03.html> (2014).

597. Weiling, F., Historical study: Johann Gregor Mendel 1822–1884. *American Journal of Medical Genetics* 40: 1–25 (1991).

598. Hamilton, B. A. and Yu, B. D., Modifier Genes and the Plasticity of Genetic Networks in Mice. *PLoS Genetics* 8: e1002644 (2012).

599. Stanford, P. K., August Weismann's theory of the germ-plasm and the problem of unconceived alternatives. *History and Philosophy of the Life Sciences* 27: 163 (2005).

600. Keller, E. F., From gene action to reactive genomes, *The Journal of Physiology* 592: 2423–9 (2014).

601. Lewontin, R. C., *It Ain't Neccessarily So* (Granta, 2000), p. 138.

602. Cohen, J., The human genome, a decade later, *MIT Technology Review*, <http://m.technologyreview.com/featuredstory/422140/the-human-genome-a-decade-later/> (2011).

603. Malik, K., The gene genie, *New Statesman*, <http://www.newstatesman.com/node/138054> (2000).

604. Brenner, S., Life sentences: Detective Rummage investigates. *The Scientist* 16: 15 (2002).

605. Alterovitz, G., Liu, J., Chow, J., and Ramoni, M. F. Automation, parallelism, and robotics for proteomics. *Proteomics* 6: 4016–22 (2006).

606. Bahassi, E. M. and Stambrook, P. J., Next-generation sequencing technologies: breaking the sound barrier of human genetics. *Mutagenesis* 29: 303–10 (2014).

607. Reardon, S., Fast genetic sequencing saves newborn lives, *Nature News*, <http://www.nature.com/news/fast-genetic-sequencing-saves-newborn-lives-1.16027> (2014).

608. Finkel, E., The trouble with genes, *Cosmos Magazine*, <http://www.npc.org.au/assets/files/documents/journalismAwards/2011%20Awards/ElizabethFinkel_TheTroubleWithGenes.pdf> (2010).

609. Van Regenmortel, M. H. V., Reductionism and complexity in molecular biology. *EMBO Reports* 5: 1016–20 (2004), p. 1016.

610. Perkel, J. M., This is your brain: mapping the connectome, *Science*, <http://www.sciencemag.org/site/products/lst_20130118.xhtml> (2013).

611. Smalheiser, N. R., The RNA-centred view of the synapse: non-coding RNAs and synaptic plasticity. *Philosophical Transactions of the Royal Society of London. Series B, Biological Sciences* 369: pii: 20130504 (2014).

612. Leybaert, L. and Sanderson, M. J., Intercellular Ca(2+) waves: mechanisms and function. *Physiological Reviews* 92: 1359–92 (2012).

613. Kopell, N. J., Gritton, H. J., Whittington, M. A., and Kramer, M. A., Beyond the connectome: the dynome. *Neuron* 83: 1319–28 (2014).

INDEX OF NAMES

Bold type is used to indicate reference to figures.

INDEX OF SUBJECTS

Bold type is used to indicate reference to figures.

ANCESTORS IN OUR GENOME

The New Science of Human Evolution

Eugene E. Harris

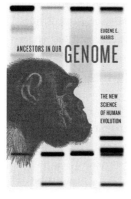

978-0-19-997803-8 | Hardback | £18.99

'Simply indispensable for any reader wishing to learn about the latest research on human origins.'
Library Journal

'In the 'Age of Genomics,' this book is an absolute must-have for anyone interested in human evolution. In the most accessible manner, Eugene E. Harris enlightens how and why genomes represent such powerful evidence to understand our past.'
Jean-Jacques Hublin, Max Planck Institute for Evolutionary Anthropology

Geneticist Eugene Harris presents us with the complete and up-to-date account of the evolution of the human genome. Written from the perspective of population genetics, *Ancestors in Our Genome* traces human origins back to their earliest source among our human ancestors, and explains some of the challenging questions that scientists are currently attempting to answer. Harris draws upon extensive experience researching primate evolution in order to deliver a lively and thorough history of human evolution.

Sign up to our quarterly e-newsletter **http://academic-preferences.oup.com/**

BIOCODE

The New Age of Genomics

Dawn Field and Neil Davies

978-0-19-968775-6 | Hardback | £16.99

'This lovely, reaching, important book shows us the front edge of a scientific movement that is transforming, simultaneously, science and our understanding of the world. If you want to understand the biological future, read this book.'

Rob Dunn

The living world runs on genomic software—what Dawn Field and Neil Davies call the 'biocode'—the sum of all DNA on Earth. Since the whole human genome was mapped in 2003, the new field of genomics has mushroomed and is now operating on an affordable, industrial scale. We can check our paternity, find out where our ancestors came from, and whether we are at risk of some diseases.

The ability to read DNA has changed how we view ourselves and understand our place in nature, and has opened up unprecedented possibilities. Already the first efforts at 'barcoding' entire ecological communities and creating 'genomic observatories' have begun. The future, the authors argue, will involve biocoding the entire planet.

FREAKS OF NATURE

And what they tell us about evolution and development

Mark S. Blumberg

978-0-19-921306-1 | Paperback | £8.99

'This book offers a unique perspective, challenging our view of science, evolution, and social archetypes by examining the nature of malformations. It would be a worthwhile addition to the library of students and scholars alike.'

Kerby C. Oberg, MD, PhD, Loma Linda University

Two-legged goats, conjoined twins, 'Cyclops' infants with a single eye in the middle of their forehead, double-headed snakes, and Laloo, a man with a partially formed twin attached to his chest ... In *Freaks of Nature*, Mark S. Blumberg turns a scientist's eye on these unusual examples of humans and other animals, showing how a subject once relegated to the sideshow can help explain some of the deepest complexities of biology.

Sign up to our quarterly e-newsletter **http://academic-preferences.oup.com/**

LIFE UNFOLDING

How the human body creates itself

Jamie A. Davies

978-0-19-967353-7 | Hardback | £20.00

'A demanding but wonder-filled account of the simple interactions that create complex structures.'

New Scientist

Where did I come from? Why do I have two arms but just one head? How is my left leg the same size as my right one? Why are the fingerprints of identical twins not identical? How did my brain learn to learn? Why must I die?

Life Unfolding tells the story of human development from egg to adult, showing how our whole understanding of how we come to be has been transformed in recent years. Highlighting how embryological knowledge is being used to understand why bodies age and fail, Jamie A. Davies explores the profound and fascinating impacts of our newfound knowledge.

Sign up to our quarterly e-newsletter **http://academic-preferences.oup.com/**

MISMATCH

The Timebomb of Lifestyle Disease

Peter Gluckman and Mark Hanson

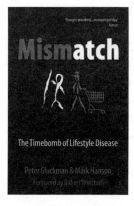

978-0-19-922838-6 | Paperback | £16.99

'Thought-provoking...this book conveys admirably, for a non-specialist reader, the implications of an important idea.'

Michael Sargent, Nature

'A fascinating and important journey through the development and evolution of human health.'

Lewis Wolpert

We have built a world that no longer fits our bodies. Our genes—selected through our evolution—and the many processes by which our development is tuned within the womb, limit our capacity to adapt to the modern urban lifestyle. There is a mismatch. We are seeing the impact of this mismatch in the explosion of diabetes, heart disease, and obesity. Bringing together the latest scientific research in evolutionary biology, development, medicine, anthropology, and ecology, Gluckman and Hanson argue that many of our problems as modern-day humans can be understood in terms of this fundamental and growing mismatch. It is an insight that we ignore at our peril.

MISSING LINKS

In search of human origins

John Reader

978-0-19-927685-1 | Hardback | £25.00

This is the story of the search for human origins—from the Middle Ages, when questions of the earth's antiquity first began to arise, through to the latest genetic discoveries that show the interrelatedness of all living creatures. John Reader's passion for this quest, and the field of palaeoanthropology, began in the 1960s when he reported for *Life Magazine* on Richard Leakey's first fossil-hunting expedition to the badlands of East Turkana, in Kenya. Drawing on both historic and recent research, he tells the fascinating story of the science as it has developed from the activities of a few dedicated individuals, into the rigorous multi-disciplinary work of today.

Sign up to our quarterly e-newsletter **http://academic-preferences.oup.com/**

NATURE'S ORACLE

The life and work of W. D. Hamilton

Ullica Segerstrale

978-0-19-860728-1 | Paperback | £16.99

'As geniuses often are, he was a complex character and an exceptional challenge for any biographer. Ullica Segerstrale is ideally qualified to rise to that challenge. She achieves a genuinely affectionate yet warts-and-all portrait of her subject, combined with a good understanding of the deep subtleties of his thinking. Those who loved him, as I did, and those who wish to know more of the astonishing originality and versatility of his contributions to science, will treasure this book.'

Richard Dawkins

W. D. Hamilton was responsible for a revolution in thinking about evolutionary biology—a revolution that changed our understanding of life itself. In this illuminating and moving biography Ullica Segerstrale documents Hamilton's extraordinary life and work, revealing a man of immense intellectual curiosity, an uncompromising truth-seeker, a naturalist and jungle explorer, a risk-taker, an unconventional scientist with a poet's soul and a deep concern for life on earth and mankind's future.

Sign up to our quarterly e-newsletter **http://academic-preferences.oup.com/**

WHAT IS LIFE?

How Chemistry Becomes Biology

Addy Pross

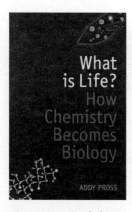

978-0-19-968777-0 | Paperback | £9.99

'I don't pretend to understand the chemistry—but by using analogies about boulders rolling down hills, and cars driving up them, Pross does a good job of explaining the principle.'

Brandon Robshaw, *Independent on Sunday*

Living things are hugely complex and have unique properties, such as self-maintenance and apparently purposeful behaviour which we do not see in inert matter. So how does chemistry give rise to biology? What could have led the first replicating molecules up such a path? Now, developments in the emerging field of 'systems chemistry' are unlocking the problem. The gulf between biology and the physical sciences is finally becoming bridged.